Risk, Reliability and Safety Engineering

In this era of globalization and competitive scenario there is a conscious effort to ensure that while meeting the reliability targets the potential risk to society is minimal and meet the acceptability criteria towards achieving long term targets, including sustainability of a given technology. The objective of reliability is not only limited to customer satisfaction but also important for design, operating systems, products, and services, while complying risk metrics. Particularly when it comes to complex systems, such as, power generation systems, process systems, transport systems, space systems, large banking and financial systems, pharmaceutical systems, the risk metrics becomes an overriding factor while designing and operating engineering systems to ensure reliability not only for mission phase but also for complete life cycle of the entity to satisfy the criteria of sustainable systems.

This book series in Risk, Reliability and Safety Engineering covers topics that deal with reliability and risk in traditional sense, that is based on probabilistic notion, the science-based approaches like physics-of-failure (PoF), fracture mechanics, prognostics and health management (PHM), dynamic probabilistic risk assessment, risk-informed, risk-based, special considerations for human factor and uncertainty, common cause failure, AI based methods for design and operations, data driven or data mining approaches to the complex systems. Within the scope of the series are monographs, professional books or graduate textbooks and edited volumes on the following topics:

- Physics of Failure approach to Reliability for Electronics
- Mechanics of Failure approach to Mechanical Systems
- Fracture Risk Assessment
- Condition Monitoring
- Risk Based-In-service Inspection
- Common Cause Failure
- Risk-based audit
- Risk-informed operations management
- Reliability Cantered Maintenance
- Human and Institutional Factors in Operations
- Human Reliability
- Reliability Data Analysis
- Prognostics and Health Management
- Risk-informed approach
- Risk-based approach
- Digital System Reliability
- Power Electronics Reliability
- Artificial Intelligence in Operations and Maintenance
- Dynamic Probabilistic Risk Assessment
- Uncertainty
- Aging Assessment & Management
- Risk and Reliability standards and Codes
- Industrial Safety

Potential authors who wish to submit a book proposal should contact:
Priya Vyas, Editor, e-mail: Priya.vyas@springer.com

Prabhakar V. Varde · Manoj Kumar ·
Mayank Agarwal
Editors

Advances in Risk-Informed Technologies

Keynote Volume (ICRESH 2024)

Springer

Editors
Prabhakar V. Varde
Reactor Group
Bhabha Atomic Research Centre
Mumbai, Maharashtra, India

Manoj Kumar
Control Instrumentation Division
Bhabha Atomic Research Centre
Mumbai, Maharashtra, India

Mayank Agarwal
Bhabha Atomic Research Centre
Mumbai, Maharashtra, India

ISSN 2731-7811 ISSN 2731-782X (electronic)
Risk, Reliability and Safety Engineering
ISBN 978-981-99-9121-1 ISBN 978-981-99-9122-8 (eBook)
https://doi.org/10.1007/978-981-99-9122-8

This Springer imprint is published by the registered company Springer Nature Singapore Pte Ltd.
The registered company address is: 152 Beach Road, #21-01/04 Gateway East, Singapore 189721, Singapore

Preface

In this evolving global world-order where apart from economics and sustainability as the two major indicators, there are overriding factors, like 'de-risking'—that covers random as well motivated or intended components of risk (safety and security)—and 'reliability' that form an integral part of dependability metrics. The current engineering systems, having their roots in technological evolution, are essentially governed by rules or specifications, best-practices, hands-on learning, experience, and peer-reviewed and enforcement systems. Here, the scope for interpretation, at times, is limited towards arriving at decisions.

We witness phenomenal progress in science and technology that is also reflected in advanced technologies that not only support but effectively enhance safety and reliability objectives. The advancement in safety and reliability technology in safety-critical systems, viz., structural, process, space, nuclear, water transport, and rail and road transport are a testimony or tribute to the success of state of the art in science and technology. However, if we look at the history of accidents and societal challenges, we still see that further efforts are required. One of the major observations has been that the human factor is one of the major contributing factors to accidents and unrest, and one still wonders—'a lot has to be done even now'.

If the present international eco-system is any indicator, then it is evident that for true progress and happiness, we need a profound value-based system where 'consciousness or to be more precise conscience-driven value system is the need of the hour'. Of course, 'quality' metrics is effective to a great extent ensuring desired performance and dependability-related aspects for engineering and social systems as a starting point. Here come risk and reliability attributes that have potential to ensure product, system and service overall performance employing quantitative goals and criteria, to a large extent, in perpetuality.

However, the challenge is the science and technology of risk and reliability is yet to achieve the higher level of considerations, e.g., demands for higher capability in terms of capturing dynamic, common cause failure, human factor and data and models with an acceptable level of uncertainty, for improved acceptability, and further become part of risk-informed and further risk-based engineering. Here, the risk also covers loss of availability and reliability that might lead to loss of production and compromised services that eventually and adversely impact the business bottom-line.

The state of the art in PRA is generally considered as matured enough to form a part of risk-informed evaluation. The reason is, PRA provides an integrated, structured and quantified model with provision to characterize uncertainty, documented required for any review and evaluation. Here, the Probabilistic Risk Assessment (PRA) provides a needed platform to either complement or supplement the existing deterministic methods as part of a risk-informed approach or as an alternate or standalone approach for not only risk assessment but also for the development of risk-based applications requiring identification and prioritization, e.g., risk-based in-service inspection, generally one-time activities—ageing management and routine maintenance management, safety significance identification to support prioritization of regulatory review management, etc. The increasing interest in the use or adoption of risk-based or risk-informed approach and effective solutions that can be seen in the open literature shows that these tools and methods will become an integral part of design, operation and regulation.

In the present context, there appears a silver lining in the context of de-risking our systems, structures and components. The phenomenal growth of advanced technologies including digital, complex computing, artificial intelligence & machine learning coupled with availability of data and experience is creating an eco-system for advanced research and development, towards addressing real-time engineering challenges. This is truly a promising development for further consolidating the effectiveness of the risk-informed approach and further development or improving risk-based technology as applicable to complex system engineering systems. The objective is to support various stages of development right from conceptualization, design, commissioning, operation, regulation and finally disposal as part of tracking sustainability.

The above discussion provides the background for identifying the title and theme of this book entitleds *Advances in Risk-Based Approach for Complex Engineering Systems—Transformative Role of Evolving Technologies*. Even though the plenary sessions of International Conference on Reliability, Safety and Hazard-2024 (ICRESH-2024) proceedings comprised seventeen keynote talks, this volume covers only 12 full-length papers. The four manuscripts were not available due to time and logistic constraints. The keynote talks in ICRESH-2024 covered a wide spectrum of topics dealing with the application of advanced tools and methods right from risk and reliability assessment, role of digital technology in general and digital twins in particular for the simulation to support, training, performance and model development, artificial intelligence and machine learning approaches as operator support aids, data-driven-based prognostics and management, etc. There is a dedicated chapter on organizational and safety technology management aspects, through an integrated approach to safety, touching upon international practices to support national regulatory objectives. Apart from this, the integrated view of asset management employing transformative technologies towards addressing Industry-4 requirements also formed a part of the conference proceedings.

It can be argued that the treatment of the subject is not exhaustive; however, the major objective of plenary sessions, i.e., to discuss the core evolving technologies and its implementations, has been met. As such, this volume can be considered to complement the contributory book volume, comprised of full-length papers as part of ICRESH-2024 proceedings. Further, pre-conference tutorial lectures provide an improved perspective on the subject, i.e., the role of reliability and risk in eliminating or reducing the potential hazard for complex engineering systems, through the development of insight on prevention, enhanced tolerance(s) and/or removal of the failure(s) or consequences of failure.

We sincerely thank our keynote speakers, many of whom have come from a long distance in India and abroad to grace the ICRESH-2024 and enlightened the participants and organizers of the event. We thank them for their support and for presenting their excellent work capturing their R&D and professional experience.

Special thanks to Springer team, Ms. Priya Vyas, Senior Editor, and colleagues for the development and management of the tasks associated with this volume and for bringing out the keynote talk book of ICRESH-2024, well in time.

Mumbai, India
November 2023

Prabhakar V. Varde
Manoj Kumar
Mayank Agarwal

Contents

1 **Tends in Engineering Asset Management: The Impact of Transformative Technologies on Risk and Reliability Management** 1
Uday Kumar

2 **Standards for Probabilistic Risk/Safety Assessments of Nuclear Power Plants and High-Level Nuclear Safety Goals—An Overview** 15
Vinod Mubayi

3 **Asset Management: A Holistic Approach to Cost Reduction, Risk Mitigation, and Performance Enhancement** 25
Gopinath Chattopadhyay

4 **Harnessing AI for Reliability and Maintenance** 33
Pierre Dersin

5 **MIRCE Science: Solar Storm as a Mechanism of Motion of Autonomously Working Systems Through MIRCE Space** 49
Jezdimir Knezevic

6 **The Development of the Integrated System Failure Analysis and Its Applications** 63
Carol Smidts and Xiaoxu Diao

7 **Digital Twins: Definition, Implementation and Applications** 79
Diego Galar and Uday Kumar

8 **Digital Twin for RAMS** 107
Bhupesh K. Lad, Ram S. Mohril, Ishika Budhiraja, and Joydeep Majumdar

9 **Software Technology Management Under Stochastic Environment** 119
P. K. Kapur and Avinash K. Shrivastava

10 Human Factors Engineering, Product Development and Sustainable Performance in Organizations: Issues and Challenges from an International Perspective 137
Pradip Kumar Ray

11 Advancements in Safety Assessment Methods and Techniques for Analysis of Internal and External Hazards 147
Janaki Devi Kompella

12 Integrated Approach to Nuclear Safety at NPCIL 157
Sameer Hajela

About the Editors

Prof. Prabhakar V. Varde started his carrier at Bhabha Atomic Research Centre in 1983 as nuclear engineering trainee of BARC Training School in 27th Batch and joined erstwhile Reactor Operations and Maintenance Group now Reactor Group and served initially as commissioning and later operations engineering for Dhruva—a 100 MW research reactor at BARC and rose through the administrative ladder and retired in 2019 as Associated Director, Reactor Group. During his service, he completed his Ph.D. from IIT, Bombay in 1996 in AI based operator advisory system and later focused his research on nuclear safety in general and Risk-based engineering in particular, while working for reactor related services responsibilities.

Alongside his regular duties he continued R&D in the area of Risk and Reliability and Academics. He also served as Senior Professor, Guide and Member of the Board of Studies in Engineering Sciences of Homi Bhabha National Institute, Mumbai. He also served as Indian specialists/experts, to International Atomic Energy Agency (IAEA), Vienna, and Nuclear Energy Agency (OECD/NEA) France. He has been on select Panel for recruitment/promotions and Ph.D. and M.Tech. Examiner at IITs and Universities.

He did his postdoctoral research at KAERI, South Korea and served as Visiting Professor at CALCE University of Maryland, USA. He has over 250 research publications at national and international level which also includes co-authored/edited 18 books, technical reports and proceedings. Recently, he has published a book entitled 'Risk-conscious Operations Management'. He received many awards and recognition. Presently, he is serving as DAE Raja Ramanna Fellow, at BARC. He is founder of (Society for Reliability & Safety) and presently serving as President, SRESA and Editor-in-Chief for SRESA's International Journal for Life Cycle Reliability and Safety.

Dr. Manoj Kumar received his B.Tech. (Electronics & Communication) from JMI, N. Delhi (1998) and Ph.D. (Engg.) from IIT Bombay (2008). He has been with Control Instrumentation Division of Bhabha Atomic Research Centre (BARC), Mumbai as Scientific Officer since 1998. He is currently working in the area development & dependability analysis of computer based systems for safety applications and prognostics of electronic systems. He is associated with BRNS (Board of Research in Nuclear Science) in its activities as project collaborator and reviewer. He is also associated with HBNI (Homi Bhabha National Institute) as guide for its post graduate programs. He has over 30 research papers to his credit and has supervised fifteen M.Tech. thesis. He has authored two books in the area of dependability modelling and accelerated life testing of electronic systems. At present he is guiding one Ph.D. candidate of HBNI in the area of prognostics of electronic systems. He is on the editorial board of three international journals and managing editor of Journal—Life Cycle Reliability and Safety Engineering. He was also the member of working group for IEEE Std1856-2017 (IEEE standard Framework for Prognostics and Health Management).

Mayank Agarwal is a mechanical engineer. After successful completion of training from 51st batch of BARC training school, he joined Bhabha Atomic Research Center, Mumbai. completed his M.Tech in Nuclear Science Engineering. As part of his M.Tech. thesis he worked on a project on development of a prototype of a Risk Monitor named as 'Risk-based Operation and Maintenance Management System'. He also worked with a team involved in developing a risk-based approach for regulatory re-licensing of a reprocessing plant. Based on the experience and research insights, he also contributed as to writing a departmental report and finally co-authored a paper on the subject in a reputed international journal.

He is working as Mechanical Maintenance Engineer for a Research Reactor at Bhabha Atomic Research Reactor. He has expertise in maintenance of Safety and Safety Related Equipments, HVAC system of nuclear facilities and Fuelling Machine.

Chapter 1
Tends in Engineering Asset Management: The Impact of Transformative Technologies on Risk and Reliability Management

Uday Kumar

Introduction

In recent years, technology has played an increasingly vital role in revolutionizing the management of risk and reliability in engineering assets, presenting both unprecedented opportunities and unforeseen challenges. However, with the advent of transformative technologies, the engineering asset management paradigm has shifted dramatically. This transformation is driven by the recognition that proactive, data-driven approaches can significantly enhance the operational performance of assets while mitigating risks and ensuring reliability. Further, these technologies empower engineering managers to make informed decisions while simultaneously minimizing operational costs. Consequently, asset managers across industries have eagerly embraced digital technologies, with the aim of achieving operational excellence. The deployment and use of industrial Internet of Things (IoT), Artificial Intelligence (AI), Machine Learning (ML), and 5G communications, has provided the asset managers with tools for near perfect cost optimization and risk elimination. As we look into the future, it is evident that technologies will continue to evolve like living organism, providing even more advanced solutions for engineering asset management.

The convergence of AI, IoT, and 5G communication is expected to open new frontiers in real-time monitoring and predictive maintenance. Furthermore, advancements in data analytics will enable asset managers to extract deeper insights from their data, leading to more informed decision-making. For high-risk industry, RAMS (Reliability, Availability, Maintainability and Safety) engineers must get used to new tools and technologies that will reduce the risk and eliminate unforeseen reliability

U. Kumar (✉)
Luleå University of Technology, Luleå, Sweden
e-mail: uday.kumar@ltu.se

P. V. Varde et al. (eds.), *Advances in Risk-Informed Technologies*, Risk, Reliability and Safety Engineering, https://doi.org/10.1007/978-981-99-9122-8_1

and safety issues. Further, during the last two decades the area of PHM (Prognostics and Health Management) has been evolving and complementing the area of RAMS to eliminate technical and financial risks from asset operation with deployment of transformative technologies.

RAMS (Reliability, Availability, Maintainability and Safety) & PHM (Prognostics and Health Management)

RAM management traditionally focuses on reducing failure occurrences and devising cost-effective maintenance programs to mitigate associated safety, business, and societal risks, considering the entire asset lifecycle. A well-structured RAMS program has become an indispensable driver of sustainable asset management for all organizations and plays a pivotal role in adapting to the challenges posed by climate change. It holds the key to ensuring competitiveness, the safe delivery of services, and the attainment of sustainability objectives. RAM parameters are determined and dimensioned during an asset's design phase, allowing for objective testing, verification, and certification. In a broader life cycle perspective, RAM assurance programs contribute to product support strategies aimed at mitigating total business risks while accounting for the unique operational context. This phase deals with population-level considerations.

PHM is operationalized during the operational phase and revolves around individual assets or engineered objects. RAM assists in planning PHM programs during operations, while the availability of PHM programs aids in the dimensioning of RAM parameters during the design phase. The overarching goal of a PHM program is to enhance asset availability by affording early proactive maintenance planning, thereby averting the consequences of unforeseen failures. PHM programs encompass detection, diagnostics, prognostics, and the identification of the Best Possible Solution (BPS) in the presence of various sources of risks and uncertainties.

Traditional reliability techniques harness probability models to characterize failure arrival rates, identify critical failure modes, and assess their severity through programs like Failure Modes and Effects Criticality Analysis (FMECA). These insights are instrumental in developing PHM algorithms. PHM algorithms, in turn, enhance classical reliability efforts by lowering the bounds on design reliability. In essence, a well-structured PHM program or tool obviates the need for extra reliability provisions, such as redundancies or additional costs. Maintenance serves to compensate for deficiencies in reliability by leveraging maintainability characteristics, thereby reducing maintenance action durations. PHM further facilitates accurate estimation of Remaining Useful Life (RUL) and the formulation of cost-effective maintenance tasks and policies leading to minimization of total asset operational risks. Figure 1.1 illustrates the concept of RUL.

The paper aims and underscores the profound significance of digital and enabling technologies in modern engineering asset management. It explores the associated challenges and opportunities, offering insights into how these technologies can reshape the future of asset management.

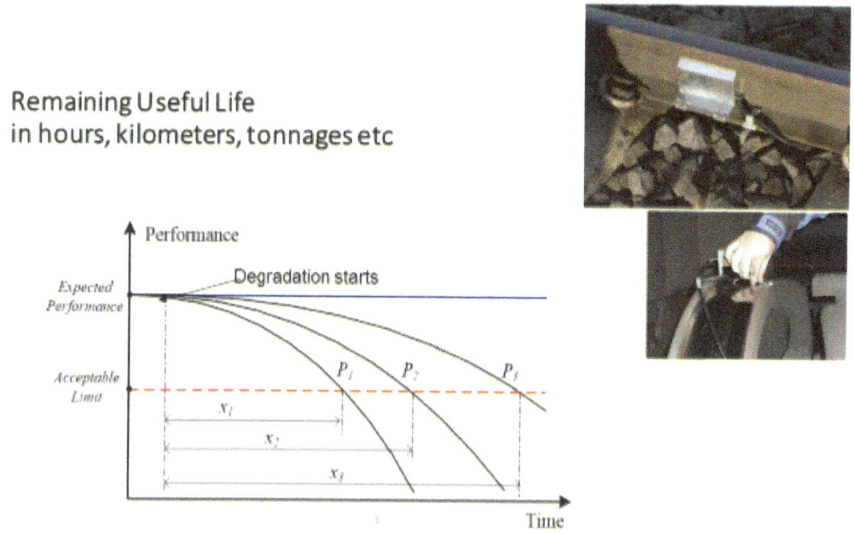

Fig. 1.1 Estimation of remaining useful life

New Technology for Asset Monitoring and Maintenance

The transformative technologies facilitate or are expected to facilitate correct decisions and actions at the lowest possible cost using the power of predictive and prescriptive analytics by the asset managers. These are expected to support organisations' digital transformation journey and operations goals. These technologies can be broadly classified as supporting, optimizing and transformative technologies, as outlined in Fig. 1.2, and collectively provide the foundation for the predictive technologies, being used for the estimation of the Remaining Useful Life (RUL) of assets. Condition monitoring, Multiphysics simulation, RAMS (Reliability, Availability, Maintainability and Safety) modelling, LCC (Life-Cycle Costs) analysis, etc. arrive at the correct optimal maintenance decision form the core of supporting and optimizing technologies. Other technologies such as Virtual Reality (VR), Augmented Reality (AR), predictive and prescriptive analytics, Industrial Internet of Things (IIoT), 5G communication technologies can offer near perfect solutions (even in real time) for the management of new or existing assets and are collectively termed as transformative technologies. For successful implementation domain knowledge is critical and expert with deep insight into the engineering system design and operation must be part of the technology deployment and implementation team.

Furthermore, for the successful development and implementation of transformative technologies and solutions, there is a spoken need for the convergence of the Operational Technology (OT) and Information Technology (IT).

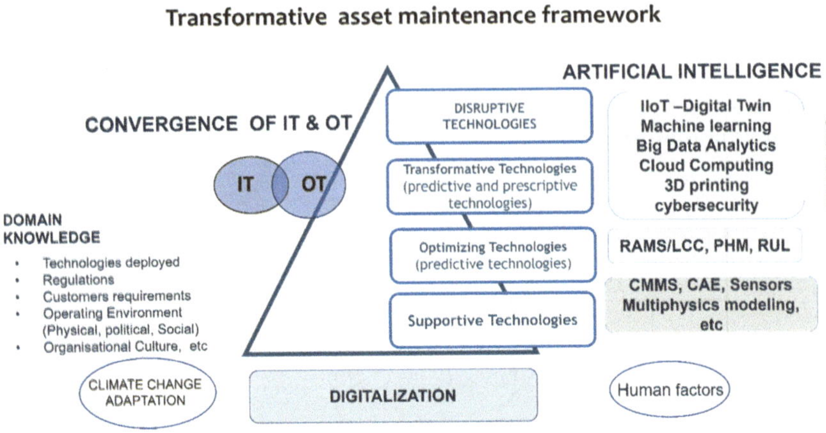

Fig. 1.2 Transformative asset maintenance framework

Digitization and Digitalization

The effort to develop and implement new technology is founded on digitalization of assets. The digitalisation process in industry and the corresponding implementation of AI technologies require availability and accessibility of both data and models. Data and models are considered digital assets that affect system's dependability during its whole lifecycle.

It provides directions for digitization, digitalization and digital transformation of maintenance process.

- Digitization is the process of converting continuous information to digital format. It is the conversion of information (i.e., objects, images, sounds, documents, data, etc.) into an electronically stored digital format that can be accessed with right tools, authorization and infrastructures.
- Digitalization is the process of leveraging value hidden in the data to improve business processes. Digitalization facilitates use of digital technologies and data to visualize the physical health of the asset and develop optimal solution for the operation and maintenance.

In short Digitization encompasses technologies that are aimed to enable the process of transforming analogue data to digital data. Digitalisation on the other hand encompassed to the provision of digital services which creates value to its user.

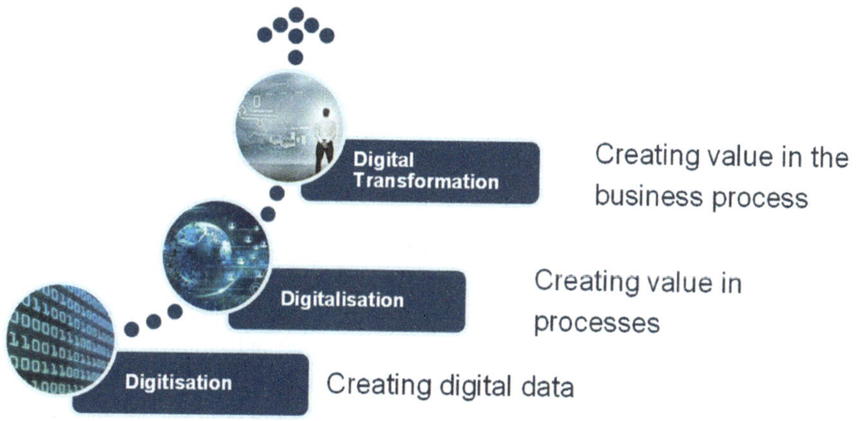

Fig. 1.3 Digitalization and digital transformation (Kumari 2022)

Digital Transformation

Digital transformation is the transformation of business activities, processes, products, and models by use of digital technologies to create added value in the business process. The main goal is to facilitate correct decision making considering several contextual variables to optimize maintenance decisions considering total business risk, as represented in Fig. 1.3. Digital transformation is enabling asset managers to keep pace with evolving customer requirements to customized business solution by integrating changing needs in real time without deploying additional resources.

Technology Roadmap and Operational Risks

Organization's Technology Roadmap is a high-level visual plan which shows the organization's technology strategy in relation to corporate strategy, technology procurement, development and implementation plan with milestones and required resources. The main goal of such road map is to increase the system dependability, reduce life cycle costs of system and finally reduce the operational risks leading to elimination of business risks. The technology roadmap exploits the data driven approach for mapping of work processes and for identification of best available technologies for deployment and use.

Data Driven Decision

As new technologies are becoming cheap and easy to use, there is a visible trend that asset managers are adopting data driven approach for effective operation and maintenance of assets. Data-driven approach is often associated with the processing a vast amount of data. In industrial context, it is often assumed that necessary data is acquired and available, but in practice the availability of high quality and right type of data and measurements are limited. In order to overcome the lack of data for analytics, augmentation techniques have been developed.

Data augmentation refers to techniques that are aimed to generate new synthetic datasets based on original datasets, e.g., by copying and slightly modifying or enriching the features in a dataset. For example, a coloured photo can be augmented by generating a grey-scaled-version of the same photo, and both the photos can then be fed to the AI-engine to increase its recognition capability. These techniques are commonly used in learning phase of AI (Karim et al. 2023).

The data driven approach to asset management approach can be divided into three sub parts namely (i) data acquisition (ii) information extraction (iii) action. Automation and integration of each of these stages can reduce the time and cost aspects associated with asset operation.

Automation of Data Acquisition

Automation of data collection has been in focus for a few decades. Starting from reactive systems built on PLC's which would discard the data at the end of each cycle, progress in data storage and processing techniques has paved the way for analyzing the acquired data to create a long-term view of the assets. This stored data can be in various formats depending on the asset scenario and condition monitoring requirements. Tabular and unformatted text, databases, images, and video are some of the data formats regularly used. These techniques have been successfully applied to a location-based data collection where the sensors are placed at a fixed location.

Automation of data collection from remote assets is possible through the use of drones and robotic dogs acting as sensor carriers, providing local storage and computation and finally relaying the data to each other or to central storage.

Quad copters or flying drones have become commonplace in last decade. Drones act as a stable platform as sensor carriers with sufficient load capacity and long flight times (depending on power source). Drones can operate in varying weather conditions and diverse locations such as high altitude and underground tunnels. Costs of such drones are well within reach compared to on ground drones such as robotic quadrupeds (available from Boston dynamics, ghost robotics and others) and wheel/tracks-based robots can act as sensor and computation device carriers. On ground drones compliment the flying drones by providing a high-resolution view near the ground surface. They can enter and traverse surfaces suitable for human access such as industrial complexes and vehicles with possible additions to open doors and safely work around humans. Currently on ground drones are limited in range due to limitation of battery capacity and requirement of charging base.

In case of fixed assets such as railway bridges and tunnels flying drones are for surveying and mapping purposes. They can generate accurate 3D models of during construction process and later for automation of inspection. Drones equipped with thermal or multispectral cameras can identify variations in temperature or chemical composition, aiding in identifying potential issues like equipment malfunctions or leakages. Flying drones are also capable of autonomous flight in restrictive areas with non-uniform surfaces.

During the last few years, asset managers are increasingly deploying robot dogs (see Fig. 1.4) and UAV drones and digital twins to manage their assets effectively and efficiently. Automation, Robotics and AI are key drivers for transforming the way assets are monitored and managed. The deployment of robots and application of digital twins have changed the way operational failures and maintenance tasks are identified, corrective measures are planned and executed by assets managers.

Robot dogs are increasingly being used for inspection of vehicles and, in some cases, for inspection of tracks. Robot dogs are designed to be able to move up and down stairs, unlike mobile wheeled robots and have replaced human for inspection of vehicles and infrastructures. These robots would not only localize faults, but also do some initial maintenance actions such as lubrication or tightening bolts and, hence, the overall inspection and maintenance downtimes will be significantly reduced.

On ground, drones or robotic dogs can navigate through hazardous areas, such as underground tunnels or unstable terrains, to perform inspections (see Fig. 1.5). Equipped with sensors, they can monitor air quality, detect gas leaks, and assess structural integrity. They can also carry specialized tools for minor repairs or sample collection.

Fig. 1.4 Training of robot dog for inspection of Track (Kumar 2022)

Fig. 1.5 Robot dog inspecting a roof collapse site in a mine (courtesy LKAB)

Flying drones have recently gained popularity for inspection of infrastructure (bridges). Images are usually processed from the front camera of the drones. Fast image processing and analysis will be used soon in drone-based track inspection. A combination of dogs and drones will enhance the capability of digital twins in a day-to-day management of asset operation and maintenance.

In case of offshore oil rigs flying drones are extensively used in inspections to assess the condition of tall structures, such as drilling rigs or flare stacks. They can capture detailed imagery and video footage to identify equipment wear and tear, structural damage, or potential safety hazards. Drones equipped with gas sensors can detect gas leaks and help prevent accidents. On ground drones specifically robotic dogs can be deployed on offshore oil rigs to conduct inspections in areas that are difficult or dangerous for humans to access. They can inspect equipment, pipelines, and structural components for signs of corrosion, leaks, or other anomalies.

Synchronization among air and ground drones creates a scenario for automation of data collection process. Flying drones can piggyback on ground drones and provide data acquisition coverage of larger areas during flight and provide initial reconnaissance, and survey the area and provide potential inspection points which can act as waypoints for traversal for robotic dog to carry out close-up inspection by navigating the terrain and reaching locations where the drones do not have access. In short, dogs and drones can be combined for missions to create a powerful inspection and maintenance system by leveraging the unique capabilities of each of the robots (Galar et al. 2021).

Synthetic Data

Data generated artificially without actual observations or measurements which mimics the real-world data is called as synthetic data. Various mathematical and computational techniques which simulate the statistical properties, characteristics and patterns of real data set are used for generating synthetic data. Traditionally, statistical methods have been used for generating synthetic data. Statistical methods can generate synthetic data by modeling the statistical properties of the original data set. Modern generative methods try to learn and capture the underlying data distribution to generate statistically similar data while capturing the complex patterns and structures present in the real data.

Statistical methods are suitable when data can be modeled to well understood distributions however intricate patterns in the data are lost. Generative methods can capture complex relationships within the data without requiring expandability of the underlying patterns and hence, they can closely resemble the structure and patterns in the original data.

Statistical methods dependent on a sample may not be able to follow such trends. Generative methods utilize feedback loop to evaluate the quality of the data generated. This feedback loop is formed between a data generator and a data discriminator. The data discriminator has access to real world data samples. The discriminator evaluates and gives feedback to the generator to improve the data quality. Both generator and discriminator improve over a period of time hence improving the quality of the data generated. By providing access to real world data samples continually improves the discriminator, hence the generated data can follow the trends from the real-world data samples.

Synthetic data is marred by various limitations such as requirement of large, diverse and good quality relevant datasets for the learning process. During the learning process feature engineering and outlier detection and their removal is required to improve the quality of the dataset. Finally, data validation and testing are required.

The greatest limitation of any method for generation of synthetic data is the lack of knowledge of the external state of the world. Hence synthetic data generation methods will always have the limitation if they are designed as stand-alone operator.

Digital Twins

Digital Twins are defined as a digital representation of a physical asset, system or process designed to detect, prevent, predict, control, and optimize through real time analytics to deliver maintenance engineering and business solution. Digital twins have been classified based on various factors; however classification based on data flow is highly relevant due to current technology and state of development of digital twins (Kritzinger et al. 2018) see Fig. 1.6. Digital twins require two-way data

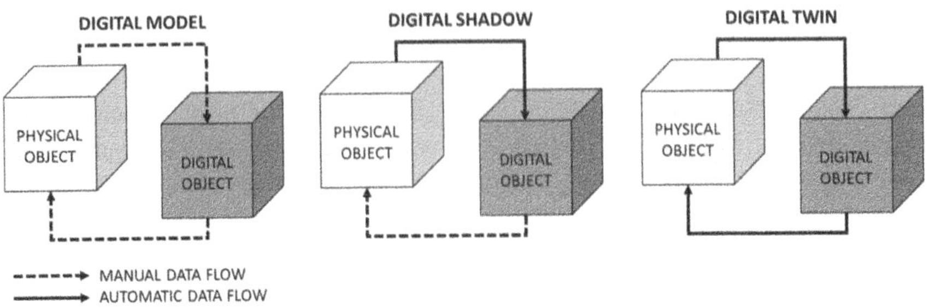

Fig. 1.6 Digital twin maturity based on data flow (Kritzinger et al. 2018)

transfer between the physical and the digital object, the classification based on this data transfer is significant since many physical systems cannot be updated by data transfer alone. The digital model depends on manual data transfer both ways this is the traditional method of simulation and modelling. Digital shadow transfers data automatically from physical object to the digital object but not the other way around, many systems developed today fall under this category. Finally, digital twin supports automatic two-way data transfer. Large number of industrial equipment support a panel for human machine interface but do not support any network interface, hence limiting the possibilities of automation. Similarly, many maintenance tasks require liveware (human) in addition to hardware and software to perform the task, usage of the term "digital twin" in such cases is debatable.

Future Trends and Directions

Asset Maintenance and Metaverse

Metaverse is a virtual shared space existing only in the digital realm, it may comprise of interconnected 3D virtual environments, augmented reality (AR) and virtual reality (VR) simulations. It allows users to interact with each other and digital objects in real time.

VR systems allow creation of complete virtual worlds for users to interact with, while AR adds virtual objects to the real world and most importantly the metaverse in not limited by the laws of physics and is independent of gravity, friction, heat etc.

Metaverse can have major impact on various aspects of maintenance, such as inspection of wheels etc. (see Fig. 1.7). Presently the inspection in railways specifically for linear assets such as overhead centenary and rail tracks is highly dependent on presence of humans and heavy equipment on the location; this is costly, dangerous and consumes track time from scheduled traffic.

Automated data captured through video and LiDAR, can be used to create digital twins of the assets. These digital twins can perform condition monitoring and predict

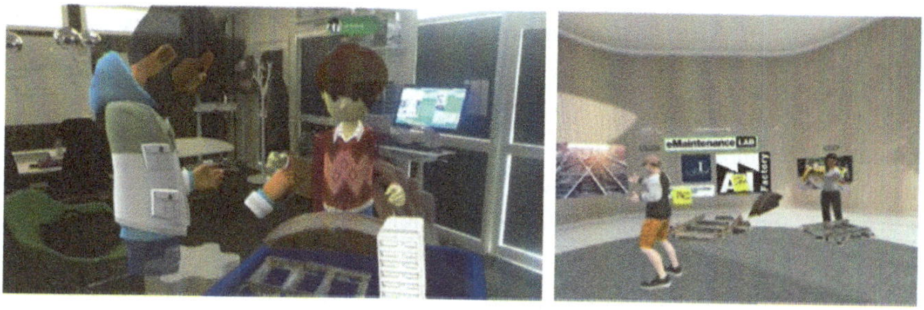

Fig. 1.7 Inspection of wheel and track in metaverse (Kour et al. 2022)

the state of the system based on system model, input data, domain knowledge, simulations, and external data streams such as weather forecasts (Patwardhan 2022).

Another common issue faced is during the training phase for new personnel is availability of various possible scenarios for training, VR allows recording/creating of environments with such scenarios for training purposes. Also, evaluation of training to assess the learning from the process can be created within such environments.

Finally, the metaverse will allow inspection to be performed by the experts, for the locations deemed to be vulnerable without stepping out of their office space. This can result in huge savings in terms of time, cost, and proper utilization of expertise. Such virtual environments do not remove the requirement of being present on site to perform the actual inspection and diagnostics related activities, AR systems can bring in digital objects to the real world, such as virtual dashboards showing inspection dates and comments next to the real-world asset.

The purpose is to facilitate seamless integration of data from on-board sensors (monitoring track condition, wayside sensors monitoring trains physical status through monitoring the temperature vibration) and ERP of the operating companies providing status of the available resources, customer requirements in real time so that correct and most appropriate decision regarding repair and maintenance action can be taken (even in real time). The AI platform essentially includes big data analytics, considering different operating environment such as moisture, temperature, snow fall and rain.

As we look to the future, it is evident that transformative technologies will continue to evolve, providing even more advanced solutions for engineering asset management. The convergence of AI, IoT, and 5G communication is expected to open new frontiers in real-time monitoring and predictive maintenance. Furthermore, advancements in data analytics will enable asset managers to extract deeper insights from their data, leading to more informed decision-making.

Integration of Augmented Reality (AR) and Virtual Reality (VR)

AR and VR technologies will find increased application in engineering asset management. Maintenance technicians will be able to access real-time information and instructions through AR glasses, facilitating quicker and more accurate repairs. VR will enable immersive training and simulation for maintenance personnel, enhancing their skills and reducing human errors leading to safe and reliable operation.

Resilience and Sustainability

Climate change and environmental concerns are driving the need for assets to withstand extreme conditions. Additionally, sustainable practices, such as energy efficiency and reduced environmental impact, will be integrated into asset management strategies. New technologies are expected to provide deeper insight into physics of climate change and suggest measures for climate change adaptation for physical and digital infrastructure. It will provide road map for designing resilient infrastructure to address the risk and challenges arising out of climate change, etc. In future all the asset managers will increasingly focus on sustainability aspects of the assets and take measures to make their assets more resilient to climate change and other unforeseen disturbances.

Blockchain for Asset Data Management

Blockchain technology will gain traction in asset management to ensure data integrity and security. Asset data, including maintenance records and performance metrics, can be securely recorded on blockchain ledgers, providing a tamper-proof and transparent record of asset history.

Issues and Challenges Associated with New Technologies

With the re-emergence of AI technologies and digitalisation, the data-driven approach has become a tool for industry in general and the transport industry in particular. Today in transport sector and other industries, there are number of promises associated with the data-driven approach such as fact-based decision-making, capability of prediction of remaining useful life of assets, improved capacity, cost efficiency, and improved sustainability with respect to environment, technology and economy.

However, these promises embrace some new challenges that need to be addressed and overcome (Kumar 2021, 2022; Galar et al. 2021) Some of these are:

- *Governance*—governance of a digital infrastructure for enabling a data-driven approach refers to the aspects of organisation, processes, policies, guidelines for data ownership, etc., which regulates collaboration and cooperation of the digital community in a digitalised environment.
- *Business*—in the context of data-driven approach business refers to models and incentives that stimulate to digital transaction between involved parties in the community. These models support the agreed governance model and facilitates implementation of data-driven approach.
- *Data Democratisation*—in a digitalised environment, the democratisation refers to availability and accessibility of data and models for the "digital citizens", i.e. individuals and organisations agreed on and committed to the defined digital governance model. There is spoken need for a framework to provide guidance for data ownership.
- *Cybersecurity/Information assurance*—information assurance in data-driven approach refers to mechanisms aimed for ensuring the aspects of cybersecurity, digital safety, and resilience of the data-driven assets, e.g. models and datasets. However, cybersecurity is weakest link in digitalisation of transport system.
- *Integration*—distributing digital assets needs to be supporting by smart integration. Integration in data-driven approach refers to mechanisms aimed for orchestration, fusion, and integration of digital assets, e.g. datasets and models.
- *Quality*—quality in data-driven approach refers mainly to measurement of aspects related quality-of-services, quality-of-data, quality-of-model. Meaning and explaining the precision of data-driven components are highly important for acceptance and fidelity models.
- Standardization: Without standardization, it is difficult to implement digital transformation in railway sector as it involves many actors with different background.
- Organizational Culture: In implementing new technologies, often organisational culture proves to a barrier that needs a special focus.
- Demographic issues, etc.

Apart from these challenges companies need to invest in their human capital and building organizational culture, to have a workplace that encourages capability-building and collaboration to spawn new, breakthrough concepts.

Concluding Remarks

In summary, transformative technologies have ushered in an era of unprecedented possibilities in engineering asset management leading to reliable and safe operation of engineering assets. By embracing digitalization, using data-driven decision-making, and deploying transformative technologies, modern day asset managers can steer their organizations toward effective management resulting in improved performance, increased operational reliability, and sustained operational excellence and substantial reduction in operational risks. However, the journey towards digital transformation necessitates the addressal of a several challenges, including technological hurdles, business process realignment, organisational, cultural and governance-related issues etc. to guarantee the continued safe, reliable, and sustainable operation of engineering assets.

References

Galar D, Kumar U, Seneviratne D (2021) Robots, drones UAVs, and UAGs, for operation and maintenance. CRC Press, Florida

Karim R, Galar D, Kumar U (2023) AI factory: theories, applications and case studies. CRC Press, Florida

Kour R, Karim R, Patwardhan A, Kumar M (2022) Metaverse for intelligent asset management. International conference on maintenance and intelligent asset management (ICMIAM), India, IEEE, 2022, s. 1–6

Kritzinger W, Karner M, Traar G, Henjes J, Sihn W (2018) Digital Twin in manufacturing: a categorical literature review and classification. IFAC-Papers Online 51(11):1016–1022

Kumar U (2021) Data-driven decisions for maintenance and life extension: keynote talk. In: Workshop on data driven approaches to building, maintaining and extending the safe life of transportation infrastructure, corporate partnership board, international transport forum, OECD, Paris, 25 February 2021

Kumar U (2022) Transformative maintenance technology for railway assets: issues, challenges and future direction: opening keynote. In: Proceedings of the fifth international conference on railway technology: research, development and maintenance, 22–25 August, 2022, Montpellier, France, Elsevier

Kumari J (2022) Augmented asset management of railway system empowered by industrial AI. Licentiate thesis, Luleå University of Technology, Luleå

Patwardhan A (2022) Enablement of digital twins for railway overhead catenary system. Licentiate Thesis, Luleå University of Technology, Luleå

Chapter 2
Standards for Probabilistic Risk/Safety Assessments of Nuclear Power Plants and High-Level Nuclear Safety Goals—An Overview

Vinod Mubayi

Introduction

Probabilistic Safety/Risk Assessment (PSA/PRA) is a widely accepted technology for investigating the risks posed by hazardous facilities. In the case of commercial nuclear power plants, the use of PSA/PRA for exploring and estimating risk goes back almost 50 years to 1975 when the pioneering Reactor Safety Study (RSS) known as WASH-1400 (US NRC 1975) carried out a quantitative risk evaluation of light-water reactors (LWRs) in the US, including pressurized water reactors (PWRs) and boiling water reactors (BWRs). The insights into the multifarious factors influencing plant risk that were revealed in the RSS helped to bring risk assessment technology closer to the regulatory arena. The results and analyses of WASH-1400 were considerably enhanced by the study of severe accident risks in PWRs and BWRs with different containment designs (large volume containments, pressure suppression containments, and ice condenser containments) carried out by the US Nuclear Regulatory Commission (NRC) in 1990 (US NRC 1990).

In 1995, the US Nuclear Regulatory Commission issued the PRA Policy Statement (US NRC 1995) that stated:

The use of PRA technology should be increased in all regulatory matters to the extent supported by the state of the art in PRA methods and data, and in a manner that complements the NRC's deterministic approach and supports the NRC's traditional defense-in-depth philosophy.

The objective of the PRA policy statement was to ensure that the applications of PRA in the nuclear industry were implemented in a consistent and predictable manner that would promote the stability and efficiency of regulatory decisions. In

V. Mubayi (✉)
Brookhaven National Laboratory (Retired), Upton, NY, USA
e-mail: vinodmubayi@gmail.com

P. V. Varde et al. (eds.), *Advances in Risk-Informed Technologies*, Risk, Reliability and Safety Engineering, https://doi.org/10.1007/978-981-99-9122-8_2

addition, the policy statement directed that the agency should use PRA and associated analyses (e.g., sensitivity studies, uncertainty analyses, and importance measures) in regulatory matters, where practical within the bounds of the state-of-the-art, to reduce risks wherever possible as well as unnecessary conservatism that was associated with deterministic regulatory requirements.

Development of PRA Standards

The increased use of PRAs/PSAs by the industry as well as in the regulatory decision-making process required that PRAs/PSAs be formulated and conducted in an accurate and consistent manner. This meant that that the quality, scope, methodology, and data used in PRAs/PSAs meet certain minimum performance standards. To achieve this objective, professional societies, industry, and the staff have undertaken initiatives to develop national consensus standards and guidance on the use of PRA in regulatory decision-making (US NRC 2020). The American Nuclear Society (ANS) Standards Board and the American Society of Mechanical Engineers (ASME) Board on Nuclear Codes and Standards (BNCS) mutually agreed in 2004 to form a Nuclear Risk Management Coordinating Committee (NRMCC). This committee was chartered to coordinate and harmonize standards activities related to probabilistic risk assessment (PRA) between the two standards developing organizations (SDOs). A key activity resulting from the NRMCC was direction to the ASME/ANS Joint Committee on Nuclear Risk Management (JCNRM) to develop PRA standards for commercial nuclear power plants that would be acceptable to the nuclear utilities, and the regulator, NRC, and structured around the three Levels of PRA for LWRs (i.e., Level 1, Level 2, Level 3) to be jointly issued by the two societies.

A PRA of an LWR is conventionally divided into three phases: Level 1 PRA carries out the analysis of the accident from the initiating event until the onset of core damage. Level 2 PRA focuses on the probabilistic treatment of accident progression from the release of core fission products to their transport in the containment and potentially from the containment to the environment. Level 3 PRA analyses the atmospheric transport of the released plume beyond the plant boundary, depletion, and the potential radiation exposure of offsite individuals through different pathways such as cloudshine. groundshine, inhalation, etc., and their health impacts such as early fatalities or latent cancers taking protective actions like evacuation and sheltering into account. Level 3 PRA also calculates the potential economic impacts of accident releases due to possible long-term relocation of affected populations as well as decontamination and/or interdiction of land and property.

The Standard for Level 1/Large Early Release Frequency Probabilistic Risk Assessment for Nuclear Power Plant Applications (ASME 2022), has been approved and published. (Large Early Release Frequency, LERF, is the frequency of severe, unmitigated, accident releases that occur early in a time frame before protective actions of the offsite population like evacuation can be effectively implemented so there is a potential for early health effects.)

The Severe Accident Progression and Radiological Release (Level 2) PRA Standard for Nuclear Power Plant Applications for Light Water Reactors (LWRs) was approved earlier for trial use and pilot application (ASME 2014). A final version is expected to be published in the near future.

The Standard for Radiological Accident Offsite Consequence Analysis (Level 3 PRA) to Support Nuclear Installation Applications was also approved earlier for trial use and pilot application (ASME 2017). Revisions to the standard are currently ongoing and a final version is expected to be published in the near future.

National consensus PRA standards are aimed at providing a set of minimum requirements that should be met, for a PRA of a plant to be considered acceptable. These standards include technical requirements of the various elements of a PRA that are focused on *what* is needed to perform that element in an acceptable way rather than *how to* perform that element. The process-related requirements address maintenance, upgrades and peer review of the standard and PRA configuration control. The peer review determines whether a PRA meets the requirements of the PRA standard.

Level 3 PRA Standard

The following material is based on the draft Level 3 PRA standard (ASME 2017) and indicates its scope and content.

The Level 3 Standard sets forth requirements for the probabilistic consequence analysis portion of PRAs/PSAs used to support risk-informed decisions for accidents involving the release of radioactive materials into the atmosphere. This portion of a PRA/PSA is typically known as a Level 3 analysis. This Standard also sets forth requirements for risk estimation based on combining the results of the Level 1 and Level 2 (Level 1/2) PRA portions (e.g., release frequencies, release characterizations) and the results of the Level 3 consequence analysis for a Level 3 PRA.

This Standard contains a brief description of each major requirement to perform a consequence analysis, and explains why it is necessary, what information results, and how it is to be used. The technical requirements for the various technical elements of a consequence analysis include (1) transport and dispersion in the atmosphere; (2) deposition processes; (3) processes that lead to the accumulation of radiation doses; (4) protective measures, such as evacuation, that can reduce radiation doses; (5) the effects of radiation doses on the human body; and (6) economic impacts. A section is also included describing how the combined risk results of a Level 1, 2, and 3 PRA can be presented. This process is referred to as "risk estimation."

A Level 3 analysis incorporates information including demography, emergency planning, physical properties of radionuclides, meteorology, atmospheric dispersion and transport, health physics, and other disciplines. Use of this information is detailed in the Level 3 Standard.

Probabilistic consequence modeling can be defined as a set of calculations of the ranges of potential adverse impacts (i.e., risk, in terms of probabilities of occurrence and magnitudes) that would follow from the dose received by humans due to a release of radionuclides. These adverse impacts, commonly referred to as "public risks," include (1) early fatalities, (2) latent cancer fatalities, (3) early injuries, and (4) non-fatal cancers. In addition, adverse impacts can occur due to contamination of property, land, and surface water. Consequence analyses may also include assessments of the economic impact of dose avoidance strategies, such as relocation of population, land and property decontamination, and interdiction of foodstuffs.

Probabilistic consequence modeling provides the means for relating these risks to the characteristics of the radioactive release and has many actual or potential applications, including the following examples:

(a) risk evaluation, generic or facility-specific, individual or the general population,
(b) environmental impact assessment,
(c) rulemaking and regulatory procedures,
(d) emergency response,
(e) development of criteria for the acceptability of risk,
(f) instrumentation needs and dose assessment,
(g) facility siting,
(h) comparison with safety goals evaluation,
(i) evaluation of alternative design features (e.g., severe accident mitigation alternatives (SAMAs) analysis), and
(j) cost–benefit analyses.

This Level 3 PRA Standard supports the quantification of a wide range of consequences/impacts to the public and the environment in the form of conditional consequences, given the postulated release(s) and risk when conditional consequence results are combined with the frequency results of the Level 1 PRA and the Level 2 PRA. Examples of the potential Level 3 consequence metrics include:

- Total effective dose (e.g., individual at the site boundary, population dose within a specified distance)
- Specific organ dose (e.g., thyroid, lung)
- Early health effects (e.g., radiological injuries, radiological fatalities)
- Latent health effects (e.g., cancers)
- Land contamination levels (e.g., area exceeding a specific Cs-137 activity)
- Economic impacts (e.g., evacuation costs, economic disruption costs, remediation costs)
- Risk values when individual consequence results are combined with release frequencies (e.g., early individual fatality risk and LCF risk for comparison to the NRC QHOs).

While the primary use of the Level 3 PRA Standard is most likely to be for LWRs, the methodology is generally applicable to any type of radioactive material released to the atmosphere from other nuclear facilities such as research reactors and fuel cycle facilities for which the release characteristics can be defined. There may be specific facilities and applications, however, where the source term phenomenology and atmospheric dispersion are complex such as releases of dense and/or reactive gases (e.g., uranium hexafluoride) that can have complex release and transport characteristics. In these cases, supplemental requirements may be needed to ensure technical acceptability.

Consequences covered within the scope of this Standard include radiation dose and induced health effects, and economic impacts, taking into account atmospheric transport and dispersion, demography, dosimetry, exposure pathways, and plant/site characteristics. The radioactive source terms and their frequencies often are passed on from supporting Level 1/2 analyses.

This Standard contains the requirements for the following technical elements of a consequence analysis: (1) radionuclide release characterization; (2) protective actions; (3) meteorological data; (4) atmospheric transport and dispersion; (5) dosimetry; (6) health effects; (7) economic factors; (8) conditional consequence quantification; and (9) risk estimation.

The requirements of each technical element may be defined at two different levels: (1) high-level requirements (HLRs), that capture the overall objective of each element and (2) supporting requirements (SRs) that are defined for each HLR and are the minimal requirements needed to satisfy the HLR The HLRs generally address attributes of the PRA technical elements, such as (a) scope and level of detail, (b) model fidelity and realism, and (c) output or quantitative results.

Objectives were established for each technical element used to characterize the respective scope of a consequence analysis. The objectives reflect substantial experience accumulated with consequence assessment development and usage. These objectives form the basis for development of the HLRs for each technical element, which were used in turn to define the SRs. The SRs are generally divided into two Capability Categories (CC), CCI and CC II. The intent of the delineation of the Capability Categories within the SRs is generally that the degree of facility specificity and the degree of realism increases from CC I to CC II. The choice of which CC to use and which SRs apply in a specific portion of a PRA depends on the application and needs a professional judgment of the analyst.

Nine technical elements are addressed in the Level 3 PRA standard. A very brief summary description of HLRs and SRs is appended under each technical element below. Detailed descriptions of the objectives, the HLRs and the SRs under each CC for each technical element are provided in Reference 7.

(a) radionuclide release characterization for Level 3 analysis (RE)

The characteristics of the radionuclide release come from the Level 1 and mainly Level 2 analyses. They include: the release categories and their binning, the specifics of the source term of each release category (i.e., the quantity of various radionuclides released, the energy and height of the release and the timing of the release), the warning time for each release category, and the aerosol particle size of each release. The frequencies of the release categories come from the Level 1 and Level 2 analyses.

(b) protective action parameters and facility and regional data (PA)

The modeling of protective actions of the offsite population is based on criteria appropriate to the phase of the accident: in the early (or emergency) phase—the first hours or days of an incident—decisions regarding evacuation and/or sheltering of the offsite population are made and implemented based on facility status and anticipated or in-progress releases, in the intermediate phase–weeks to months after release—protective actions are based on environmental measurements of contamination, and in the late/long-term phase—the subsequent months to years following a release-recovery/remediation actions such as land/property decontamination are conducted and completed, and land/property is either released for unrestricted use or condemned for habitation. Parameters such as evacuation speed and shielding factors need to be based on site-specific studies of evacuation time estimates and building stock materials, and foodstuff interdiction criteria as well as contaminated land habitability criteria on recommendations of recognized official agencies. Other facility and regional data needed include facility physical characteristics (building dimensions, stack heights). Population distributions around the facility and the region, and regional land use data (fraction of land that is water, fraction that is farmland, agricultural production).

(c) meteorological data (ME)

Accurate meteorological data at and in the vicinity of the facility are important and should cover at least 90% of hourly data over a period of a year on windspeed, wind direction, precipitation and measurement of temperature difference with height or wind direction standard deviation or observations from representative weather stations that can be used to determine the atmospheric stability class. Met data should be collected under a qualified scheme of calibration, maintenance activities, and instrument exposures. Data from recognized sources on seasonal morning and afternoon mixing heights in the region need to be compiled.

(d) atmospheric transport and dispersion (AD)

The objectives of the atmospheric transport and dispersion technical element are to develop, use, and document an atmospheric transport and dispersion (ATD) model in such a way that: conditions at the facility are represented, site specific meteorological data is used, facility and accident specific attributes are accounted for, temporal and spatial changes in meteorological conditions are considered, and deposition of radionuclide particles is included. The requirement is to ensure that an appropriate dispersion methodology is adopted that incorporates the meteorological data to determine the airborne concentration and ground deposition for input into dose models.

(e) dosimetry (DO)

The dosimetry technical element uses appropriate dose conversion factors along with the computed radionuclide concentrations and surface depositions to determine the doses received by the tissues and organs of interest due to exposure to radioactive material via each of the relevant dose pathways in such a way that applicable exposure pathways and protective action impacts are included, and appropriate dose conversion factors are used. The plume concentrations and deposition resulting from the ATD model are used to calculate doses over the exposure period(s). The analysis includes applicable exposure pathways including cloudshine, groundshine, skin deposition, skin absorption, inhalation and ingestion, and takes into account the effect of protective actions on received dose. The calculation of groundshine dose integrates the dose over the exposure time period(s), accounting for deposited materials both during and after plume passage. Acute and committed doses from modeled pathways are calculated to obtain effective dose and specific organ doses for which health effects are to be estimated.

(f) health effects (HE)

This technical element estimates the health effects of interest, such as early fatalities, early injuries, latent cancer fatalities and non-fatal cancers based on the computed doses and appropriate risk factors in such a way that health effect modeling accounts for both dose and dose rate, using parameter values from recognized sources.

(g) economic factors (EC)

This technical element ensures that the economic factors determined for the analysis use appropriate models and facility-specific and regional data in such a way that economic model parameters are clearly defined, and parameter estimates have an appropriate basis. Economic cost factors include: evacuation costs, evacuation costs, relocation costs including temporary unemployment, land value, depreciation, crop losses, decontamination costs, loss of use of offsite property, and public health costs (e.g., based on monetizing population dose).

(h) conditional consequence quantification and reporting (QT)

Conditional consequences include metrics of interest such as doses, early fatalities, latent cancers, costs, etc. that are identified and quantified in this technical element.

(i) risk estimation (RI)

This technical element identifies the risk metrics of interest such as population dose risk, early fatality risk, latent cancer risk, etc., and estimates them by combining the results of the Level 1, Level 2, and Level 3 analyses. It also describes how the combined risk results of a Level 1, Level 2, and Level 3 PRA can be presented.

This Standard is being developed by experts in various disciplines associated with consequence analysis such as severe accident source term modeling, meteorology, atmospheric transport, dosimetry, radiation health effects and risk estimation drawn from U.S. nuclear utilities, national laboratories, individual consultancies, and the U.S. NRC. A very limited amount of international participation has taken place in the process of developing the standard, but international users should be able to adapt the examples to their specific applications and regulatory requirements.

Safety Goals

The output of a Level 3 PRA is typically expressed through risk metrics such as the public health risks of early fatality and latent cancer fatality caused by exposure to the radionuclides released in a severe reactor accident. In 1986, the US NRC issued a Safety Goal Policy Statement that adopted probabilistic safety goals as rational objectives for the limits of severe accidents on public health risk (8). The safety goals that were adopted included two qualitative safety goals and two quantitative health objectives (QHOs). The qualitative goals expressed the Commission's expectation that members of the public residing in the vicinity of a nuclear power plant (NPP) should bear no significant additional risk from plant operation and that risks of generating electricity by a NPP should be comparable to or less than the risks of electric generation by other viable technologies. The two QHOs bear directly on the public health risks calculated in a level 3 PRA. QHO 1 refers to individual early fatality risk and states that the risk of an early fatality from a NPP accident to a biologically average individual (in terms of age and other risk factors) who resides within 1 mile of the site exclusion area boundary should be less than one-tenth of one percent of the sum of prompt fatality risks from all other accidents that the US population is generally exposed to. QHO 2 refers to individual latent cancer fatality risk and says the risk of a latent cancer fatality from a NPP accident to a biologically average individual within 10 miles of the site should be less than one-tenth of one percent of the sum of cancer fatality risks to the US population from all other causes. The numerical value of one-tenth of one percent of the background risk of either individual early fatality or latent cancer fatality represented, in the Commission's view, the notion of no significant additional risk from NPP accidents.

The results of major Level 3 PRA risk studies such as the NUREG-1150 study referred to in Reference 2 indicated that nuclear power plants in the United States satisfy the safety goal QHOs by a wide margin even taking into account uncertainties in the analysis. Two decades after NUREG-1150, the State-of-the-Art Reactor Consequence Analyses (SOARCA) study (US NRC 2012), performed by Sandia National Laboratories for the NRC, revisited the methods of analysis used in NUREG-1150 and repeated probabilistic risk assessments for two nuclear plants—Surry and Peach Bottom also evaluated in NUREG-1150—using updated consequence analysis tools and methodologies. While SOARCA did "not examine all scenarios typically considered in a probabilistic risk assessment," its results indicated that the QHOs were satisfied by even larger margins than in the NUREG-1150 study.

However, in the wake of the severe reactor accident that occurred at Fukushima, there has been a realization that individual fatality risks from radiation exposure do not constitute the only risk to which offsite populations are exposed. To protect the offsite public from exposure to the radiological materials released to the environment during a severe accident, various emergency protective action measures are employed, ranging from sheltering-in-place to evacuation followed by extended relocation, if necessary, and remediation or condemnation of contaminated land along with banning of contaminated food. At Fukushima, there were no radiation-induced early fatalities and any latent cancer fatalities from the radionuclides released in the accident are not expected to present a measurable increase over the background rate of cancer fatality in Japan. On the other hand, there were a significant number of non-radiation-related fatalities from traffic accidents during evacuation and some other causes such as premature deaths of relocated elderly patients. Other major consequences of the Fukushima accident have been the extended loss of homes and lands, the negative psychological impacts of long-term relocation, and the high costs involved with the remediation of contaminated land.

The non-radiation related risks of NPP accidents arise from the measures taken to prevent the offsite public from radiation exposure. A number of recent studies have suggested that rather than radiation-induced health effects the major impact of an NPP accident is what has been termed as societal risk, the social disruption caused by the relocation of large numbers of people. Bier et al. (2014) analyzed accidents at five sites in the US and concluded that the number of people relocated represents a viable measure of societal impact. Denning and Mubayi (2016) compared the monetized impact of NPP accidents to other societally disruptive events such as major hurricanes and observed that societal risk rather than individual health risk of radiation exposure is the dominant risk of NPP accidents.

Mubayi and Youngblood (2021) have proposed an additional safety goal based on societal risk to the existing NRC safety goals that consists of both a qualitative as well as a quantitative goal. The qualitative societal risk goal is that there should be no significant likelihood of extended relocation of a large number of people due to an NPP accident. The proposed quantitative goal is based on the recognition that it is the release of the relatively long-lived cesium isotopes in a NPP accident that is responsible for the extended relocation of the public in line with habitability criteria. Hence the proposed quantitative societal risk goal is that the mean frequency of release of cesium amounting to X% of the core inventory that can cause an extended relocation of more than one year of Y offsite persons should not exceed 1E-06 per reactor-year. The numerical values of X and Y are ultimately policy choices on the part of decision-makers. A preliminary review of the literature indicates that X% may range from 1 to −10% for most operating reactors in the US while the analysis in Ref. 10 suggests that Y may range from about 1E +04 to 1E+05 at several sites in the US.

A societal risk goal added to the existing US NRC safety goals would help to achieve a more comprehensive characterization of nuclear power plant safety.

References

ASME/ANS RA-1.2 (2014) Severe accident progression and radiological release (Level 2) PRA standard for nuclear power plant applications for light water reactors (LWRs). American Society of Mechanical Engineers/American Nuclear Society, [Draft for Trial Use and Pilot Application]

ASME/ANS RA-S-1.3 (2017) Standard for radiological accident offsite consequence analysis (Level 3 PRA) to support nuclear installation applications. American Society of Mechanical Engineers/American Nuclear Society, [Draft for Trial Use and Pilot Application]

ASME/ANS RA-S-1.1 (2022) Addenda to ASME/ANS RA-S-2008 standard for Level 1/large early release frequency probabilistic risk assessment for nuclear power plant applications. American Society of Mechanical Engineers/American Nuclear Society
Bier VM et al (2014) Development of an updated societal-risk goal for nuclear power safety. In: Proceedings of the conference on probabilistic safety assessment & management (PSAM-12), Honolulu, HI, pp 22–27
Denning R, Mubayi V (2016) Insights into the societal risk of nuclear power plant accidents. Risk Anal 37:1
Mubayi V, Youngblood R (2021) Reevaluating the current U.S. Nuclear regulatory commission's safety goals. Nuclear Technol 207(3):406–412. https://doi.org/10.1080/00295450.2020.1775452
US NRC (1975) Reactor safety study: an assessment of accident risks in U.S. Commercial nuclear power plants (WASH-1400). NUREG-75/014
US NRC (1986) Safety goals for the operations of nuclear power plants; policy statement; republication. Federal Register 51 FR 30028
US NRC (1990) Severe accident risks: an assessment for five U.S. nuclear power plants. NUREG-1150. Washington, DC
US NRC (1995) Federal Register, 60 FR 42622
US NRC (2012) State-of-the-art reactor consequence analyses project. NUREG/CR-7110
US NRC (2020) Regulatory Guide 1.200

Chapter 3
Asset Management: A Holistic Approach to Cost Reduction, Risk Mitigation, and Performance Enhancement

Gopinath Chattopadhyay

Introduction

The history of asset management traces back to terotechnology, encompassing the installation, commissioning, maintenance, replacement, and removal of plants and equipment. Initially, the focus was on maintenance and asset management, but the field has evolved to consider a broader array of assets, including financial, informational, human, and intangible assets such as knowledge and goodwill. This evolution reflects a holistic approach to balance costs, risks, and performance (Chattopadhyay 2016, 2018).

Assets, as defined by ISO55000 standards, encompass items, entities, or things with value or potential value to an organization. Asset management fundamentals revolve around value, leadership, culture, alignment with corporate objectives, and assurance that assets will perform as needed (ISO 2014). While this chapter primarily addresses physical assets, it is crucial to acknowledge the significance of other assets.

Asset management for physical assets was first introduced by Dr. Penny Burns in the 1980s (Asset Management History Project 1984; AMC 2021), and it gained international recognition with the Infrastructure Asset Management Manual published in New Zealand in 1996, later becoming the International Infrastructure Management Manual (IIMM) in 2000 (IPWEA 2021).

Professional societies like the Asset Management Council (AMC) in Australia, Institute of Public Works Engineers Australasia (IPWEA), the Institute of Asset Management (IAM) in the UK, and various global bodies have played a pivotal role in developing the body of knowledge in asset management (IAM 2021; GFMAM

G. Chattopadhyay (✉)
Centre for Smart Analytics, Institute of Innovations, Science and Sustainability, Federation University Australia, Ballarat, Australia
e-mail: g.chattopadhyay@federation.edu.au

P. V. Varde et al. (eds.), *Advances in Risk-Informed Technologies*, Risk, Reliability and Safety Engineering, https://doi.org/10.1007/978-981-99-9122-8_3

2017). The Global Forum of Maintenance and Asset Management (GFMAM) has facilitated a unified understanding of asset management across countries and the development of guidelines to address global asset management challenges consistently (Peterson 2007).

Capital-intensive industries worldwide face growing pressure due to demand growth, geographical challenges, and aging assets. Credit constraints and capital scarcity, add more pressure. The risks and costs with interaction often reveals the need for adequate resources for prevention and improvement initiatives. The key is to emphasize the value of proposed initiatives beyond just costs and benefits.

The asset management journey commences with aligning organizational needs with business objectives and continues through design, manufacturing/construction, operations, maintenance, and ultimately asset disposal in a cost-effective, reliable, safe, secure, and timely manner.

Overview of Asset Management

Asset management, according to Peterson, encompasses key concepts:

- Business goals driving decisions for asset use and care.
- Asset strategy determined by operational considerations.
- Maintenance and reliability as means, not ends.
- Intent to optimize the application of all resources, not just maintenance (Moore 1996).

Moore's perspective on asset management includes:

- Alignment of business expectations for assets, present and future.
- Understanding the assets' current condition and capabilities.
- The centrality of considering how and why assets are operated.
- Consideration of asset lifecycle, including design for capability, reliability, and ease of management.
- Implementation of asset management (Woodhouse 2008).

Asset management serves as a strategic platform to connect physical assets, their utilization, maintenance, and all other asset types. Woodhouse defines it as a set of disciplines, methods, procedures, and tools aimed at optimizing the whole-of-life business impact, encompassing costs, performance, and risk exposures related to assets (BSI 2008).

The International Infrastructure Management Manual (IIMM) outlines the development of:

- Asset Management (AM) policy.
- Organizational structure for delivering AM functions.
- Quality management processes supporting AM functions.

Publicly available specifications such as PAS55 (British Standard Institute 2008) paved the way for the international asset management standard ISO55000:2014. This series focuses on risk-based and informed decision-making to reduce costs, mitigate risks, and enhance performance throughout an asset's lifecycle, including acquisition, utilization, and disposal (BSI 2014a, b).

The ISO standard for asset management consists of three parts:

ISO 55000: Asset management—Overview, principles, and terminology.
ISO 55001: Asset management—Management systems—Requirements.
ISO 55002: Asset management—Management systems—Guidelines on the application of ISO 55001 (ISO 2018; Asset Management History Project 1984).

Asset Management, as per ISO55000, is defined as the "*coordinated activities of an organization to realize value from assets.*" It is guided by several principles, including recognizing that assets exist to provide value, the importance of people in asset value realization, and the need for understanding the organization's operating context and opportunities.

The ISO55000 series outlines what needs to be done but not how it should be done. Organizations must address the requirements according to their context and expectations. The standard covers various aspects, from leadership and policy to planning, support, operation, performance evaluation, and improvement.

The Global Forum on Maintenance and Asset Management (GFMAM) published the Asset Management Landscape, which defines the knowledge and skills needed for effective asset management. It includes areas like asset management strategy, decision-making, lifecycle delivery activities, asset knowledge enablers, organization and people enablers, risk and review (Peterson 2007).

To enhance personnel capabilities in asset management and minimize lifecycle costs (LCC), organizations must develop tools, techniques, and artifacts. These tools assist in managing asset value across procurement, operation, maintenance, and asset disposal stages.

Understanding an Asset and Its Remaining Life

Estimating the remaining life of an asset is a comprehensive, multidisciplinary activity that considers factors such as the asset's lifecycle, user needs, stakeholder demands, policy environment, governance framework, technical adequacy, market factors, and more. It aims to make sound decisions that address identified risks and their impacts on value, perform tasks at the right time and expenditure level, and achieve a balance between performance, cost, and risk. The starting point for this process is understanding the asset's failure mechanism.

Failure is the inability of an item to perform its intended function, often resulting from design, manufacturing, user, or maintenance-related issues. Failure analysis involves a systematic examination of probability, causes, and consequences of failures, including near misses. Failures can be categorized as primary, secondary, wear-out, sudden, gradual, partial, complete, catastrophic, or degradation, depending on their characteristics and impacts.

Failures occur throughout an asset's lifecycle due to various factors, including design flaws, operational wear and tear, and aging. These factors are typically represented in a bathtub failure curve, which encompasses decreasing, constant, and increasing failure rates during different asset lifecycle stages, including the early failure period, constant failure rate period, and wear-out period.

Understanding the remaining life of an asset is fundamental for decision-making and risk management. It involves evaluating the asset's current condition, estimating its future performance, identifying potential failure modes, and predicting when failure is likely to occur. Common approaches to estimating remaining life include reliability-centered maintenance (RCM), condition monitoring, and predictive analytics.

Reliability-Centered Maintenance (RCM): RCM is a systematic approach to analyzing the functions and potential failure modes of assets. It aims to determine the most appropriate maintenance strategies to maximize reliability, safety, and cost-effectiveness. RCM identifies failure modes, their consequences, and the most suitable maintenance actions, such as corrective, preventive, or predictive maintenance.

Condition Monitoring: Condition monitoring involves continuously or periodically assessing an asset's condition through various techniques, including vibration analysis, thermography, oil analysis, and ultrasound. It helps detect early signs of deterioration or impending failures, enabling timely maintenance interventions.

Predictive Analytics: Predictive analytics leverages data and machine learning algorithms to forecast asset performance and identify anomalies or patterns indicative of potential failures. It can provide valuable insights into the remaining life of assets and optimize maintenance schedules.

Risk Assessment and Mitigation

Risk assessment is a crucial aspect of asset management, as it helps organizations identify, prioritize, and manage risks that can impact asset performance, safety, and overall business objectives. Effective risk assessment enables organizations to make informed decisions regarding asset maintenance, replacement, and investment.

Key steps in the risk assessment process are summarized as follows:

Risk Identification: Identifying potential risks that could affect asset performance, safety, or reliability. Risks can stem from various sources, including technical, operational, environmental, regulatory, and financial factors.

Risk Analysis: Evaluating the likelihood and consequences of identified risks. This involves quantifying the probability of risk events occurring and assessing their potential impact on assets and the organization.
Risk Prioritization: Prioritizing risks based on their significance and potential consequences. High-priority risks that pose significant threats to asset integrity or business operations require immediate attention.
Risk Mitigation: Developing strategies and action plans to mitigate identified risks. Mitigation measures may include preventive maintenance, asset replacement, redundancy, improved operational procedures, and compliance with regulatory requirements.
Risk Monitoring: Continuously monitoring and reassessing risks to ensure that mitigation measures remain effective. Regular risk reviews and performance monitoring are essential to adapt to changing conditions and emerging risks.

Risk mitigation strategies should align with an organization's risk tolerance and asset management objectives. Risk tolerance refers to the level of risk that an organization is willing to accept or tolerate, considering its overall risk appetite and business goals.

Performance Enhancement Through Asset Management

Asset management practices are instrumental in enhancing asset performance and achieving optimal outcomes for organizations. Key aspects of performance enhancement through asset management include:

Reliability Improvement: Asset management strategies focus on enhancing asset reliability by implementing maintenance practices that minimize downtime, reduce failures, and extend asset lifecycles. This includes preventive and predictive maintenance approaches, condition monitoring, and reliability-centered maintenance (RCM).

Cost Reduction: Effective asset management can lead to significant cost savings by optimizing maintenance activities, reducing unnecessary expenditures, and extending asset lifecycles. Cost reduction strategies include minimizing unplanned downtime, optimizing maintenance schedules, and eliminating unnecessary asset replacements.
Enhanced Safety: Asset management prioritizes safety by identifying and mitigating risks associated with asset failures. Safety improvements can result from proactive maintenance practices, risk assessments, and compliance with safety regulations.
Asset Optimization: Asset management facilitates the optimization of asset utilization, ensuring that assets are used efficiently to meet organizational objectives. This includes optimizing asset performance, capacity, and resource allocation.
Data-Driven Decision-Making: Asset management relies on data collection and analysis to make informed decisions about asset maintenance, replacement, and

performance improvement. Data-driven insights enable organizations to prioritize actions that deliver the greatest value.
Regulatory Compliance: Asset management practices ensure that organizations comply with relevant regulatory requirements and standards, reducing the risk of non-compliance penalties and legal issues.
Sustainability: Asset management considers environmental and sustainability factors, helping organizations reduce their environmental footprint and adhere to sustainability goals.

Conclusion

Asset management, as defined by ISO55000 standards, is a holistic approach to optimizing the value, performance, and reliability of assets while mitigating risks and reducing costs. Effective asset management encompasses the entire asset lifecycle, from acquisition to disposal, and involves various disciplines, including maintenance, reliability, risk management, and data analytics.

Understanding an asset's remaining life is crucial for making informed decisions about maintenance, replacement, and resource allocation. Organizations can use methods such as reliability-centered maintenance (RCM), condition monitoring, and predictive analytics to estimate remaining life and optimize asset management strategies.

Risk assessment and mitigation are integral components of asset management, enabling organizations to identify, prioritize, and manage risks that can impact asset performance and business objectives. By developing effective risk mitigation strategies and monitoring risk factors, organizations can enhance asset reliability and safety. There are tangibles and intangible. Social risks associated with adverse impacts are difficult to identify and assess, There are counterproductive KPI's in many organisations where capex and opex are detached and not considered for whole of life analysis. Data is another challenge in terms of volume, quality, ownership and trust worthiness.

Ultimately, asset management practices lead to performance enhancement by improving asset reliability, reducing costs, enhancing safety, optimizing asset utilization, and supporting data-driven decision-making. Organizations that prioritize asset management can achieve long-term sustainability, compliance with regulations, and competitive advantages in their respective industries. A Holistic Approach to Cost Reduction, Risk Mitigation, and Performance Enhancement needs to start from knowing the assets, having a documented framework, asset management converging strategic asset management plan, asset management plan, asset specific plan in line with corporate objective and capex and opex balancing over the whole of life of the asset based on what vale.

References

AMC (2021) Asset management council. https://www.amcouncil.com.au/

Asset Management History Project (1984) A history of asset management. Int J Prod Res 22(1):17–42

BSI (2008) PAS 55-1:2008 Specification for the optimized management of physical infrastructure assets

BSI (2014a) ISO 55001:2014 Asset management—Management systems—Requirements

BSI (2014b) ISO 55002:2014 Asset management—Management systems—Guidelines for the application of ISO 55001

Chattopadhyay G (2016) Asset management excellence: optimizing equipment life-cycle decisions. CRC Press

Chattopadhyay G (2018) Asset management: past, present, and future. In: Handbook of maintenance management and engineering. Springer, pp 17–44

GFMAM (2017) Asset management landscape. https://www.gfmam.org/

IAM (2021) Institute of asset management. https://theiam.org/

IPWEA (2021) Institute of public works engineering Australasia. https://www.ipwea.org/

ISO (2014) ISO 55000:2014—Asset management—Overview, principles, and terminology

ISO (2018) ISO 55000:2018 Asset management—Overview, principles, and terminology

Moore S (1996) Reliability-centered maintenance: management and engineering methods for optimizing asset performance. Industrial Press Inc.

Peterson AS (2007). The philosophy of asset management. In: Uptime, vol. 1. PdM Technology, pp 5–11

Woodhouse J (2008) Asset management: the asset management handbook. The British Standards Institution

Chapter 4
Harnessing AI for Reliability and Maintenance

Pierre Dersin

Introduction

To keep the key critical systems of today's complex world operating smoothly and cost-effectively, reliability and asset management have become priorities for industry and governments. Accordingly, more than ever, close attention is paid to ensuring reliability performance and optimizing maintenance and asset management policies.

The last two decades have witnessed the triumph of the digital transformation and, in particular, the "Internet of things" which, through the combination of cost-effective sensors and efficient communication infrastructure, allows many industrial items of equipment to communicate in real time physical magnitudes that can be used to estimate their health condition.

This evolution, combined with the fast-paced development of advanced data processing algorithms ("analytics"), including the vast area called artificial intelligence (AI), has spurred the emergence of a discipline named Prognostics and Health Management (PHM). At the same time, it is revolutionizing reliability engineering.

Back in the 1940s until the 1960s, reliability engineering evolved as a full-fledged scientific discipline under the pressure of the Cold War and the space race- and, not surprisingly, many prominent actors were found in the United States (e.g., Barlow and Proschan 1965) and the Soviet Union (e.g. Gnedenko). They built on important pre-WWII work such as that of W. Weibull in Sweden (Weibull 1939), as well as the foundations of reliability-related statistics by the likes of Fisher (1922) and Cox (1972a). A body of methodologies was soon constituted—necessarily, with simplifying assumptions.

P. Dersin (✉)
Eumetry SaS, Louveciennes, France
e-mail: pierre.dersin@ltu.se

Luleå University of Technology, Luleå, Sweden

P. V. Varde et al. (eds.), *Advances in Risk-Informed Technologies*, Risk, Reliability and Safety Engineering, https://doi.org/10.1007/978-981-99-9122-8_4

The situation in those days is best characterized by scarcity of data, and primitive computation tools. Indeed the beginnings of reliability theory were contemporaneous with the first computers, which utilized vacuum tubes and occupied an entire room. As late as the late 1970s, computer programs—even in a place like MIT-still had to be typed on punch cards and brought to an intimidating "computer center" which would (in the best case) deliver the outputs the following day. Many countries still had manually operated analog telephone exchanges—so much for telecommunication. Under those circumstances, emphasis was placed on simplified models: statistical independence, stationarity, exponential time to failures, and the like. Metrics of interest addressed average characteristics at population level, such as MTTF or MTBF.

One looks back in awe at the incredible technological successes which were achieved under those conditions—including landing the first man on the Moon, in 1969. In parallel, maintenance practice has evolved from 'run to failure" purely corrective maintenance to preventive maintenance and, in the 1960s, spurred by the aeronautics industry, the RCM (Reliability-Centered Maintenance) methodology which links reliability engineering (or, more precisely, RAMS) with maintenance needs, and stresses the notion of functional maintenance, i.e., maintenance plans determined by the need to keep fulfilling the function. In that context, condition-based maintenance increasingly has become part of the landscape.

In the 1950s, the expression "artificial intelligence" (AI) appeared, with the 1956 Dartmouth seminar organized by J. Mc Carthy. In its most general definition, AI aims at replicating human reasoning automatically. The first direction of investigation was symbolic AI, whereby special languages (such as LISP) were created to manipulate symbolic logic. This approach had limited success in the form of rule-based expert systems, for medical diagnostics for instance. But the unreasonable expectations they had raised were crushed, and led to disillusionment in the 1980s.

In parallel, another approach was pursued, with the research on *machine learning*, which can be described as the field of study that gives computers the ability to learn without being explicitly programmed (see e.g., Alpaydin 2014).While, traditionally, computers are given a model and inputs, and apply the explicit instructions of the model to the inputs in order to generate outputs, instead, with machine learning, the computer is given inputs and outputs and is asked to find a model that could have generated the given outputs from the given inputs. Once it has found that model, it can then apply it to new inputs.

Today, AI today is often understood to mean ML but actually ML is just a subset of AI (Fig. 4.1).

Numerous machine learning methods have appeared, which essentially fall into two main categories: supervised learning, whereby the computer is trained on a set of 'labelled" input data and is given outputs corresponding to each input data set (for instance, the output word "cat" corresponds to input images of cats); and unsupervised learning, whereby the computer is just given input data and has to detect patterns somehow (for instance, through clustering). An important and growing area is also reinforcement learning, which provides feedback so that 'good 'decisions

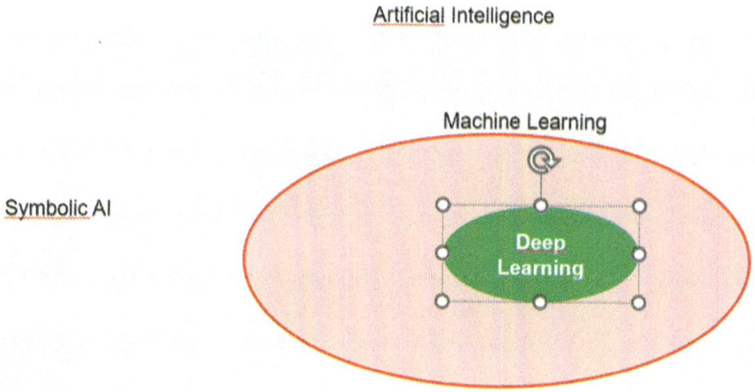

Fig. 4.1 Artificial intelligence (AI)

are rewarded and bad decisions are penalized, and in that way the algorithm learns from experience and improves over time. The catalog of ML method is huge and growing; some of the better known ones include support vector machines (SVM), KNN (K-nearest neighbors), various regression methods, random forests, ensemble methods, etc.

Generally, in PHM, signals acquired by sensors that monitor an asset are summarized by so-called 'features', which form the basis for constructing health indicators and performing anomaly detection, fault diagnostics, and if possible prognostics. Features are traditionally engineered by domain experts but, with ML, there are methods that permit automatic feature learning.

One family of methods that has had great success recently, i.e. in the last decade or so, after knowing its ups and downs since the 1950s (the "down" periods have been called "AI winters") is that of artificial neural networks (ANN) (Fig. 4.2). ANNs have been inspired by biological neural networks. They incorporate two fundamental components: neurons (represented by nodes) and synapses (represented by links). Each node computes a nonlinear function of a weighted sum of inputs. The choice of the nonlinear functions is part of the "network architecture". The values of the weights result from an optimization. For instance, in the supervised learning case, the weights are determined from a set of inputs in order to minimize some distance between the generated outputs and the target outputs. In general, a "loss function" is minimized to determine the best weights.

The ability of neural networks to learn any nonlinear function is characterized by "universal approximation theorems" (Hornik et al. 1989). This echoes somehow the intuition of MIT mathematician Norbert Wiener who, in one of his epochal books (Wiener 1964), wrote that "a learning machine operates with nonlinear feedback".

When the neural network contains more than one inner layer, it is called a deep network, and machine learning using such a network is called *deep learning*. Various neural network architectures have been introduced and used successfully. For instance the convolutional neural network (CNN) is ideal for image processing

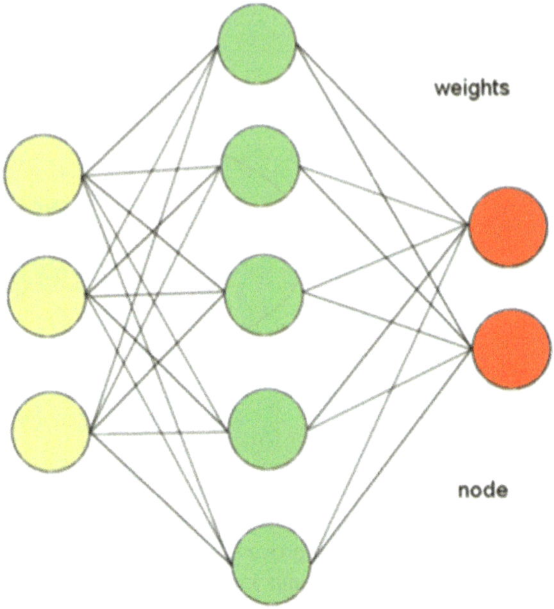

Fig. 4.2 Artificial neural network

in grids. Recurrent neural networks (RNN) include feedback loops and are quite suitable for automatic language translation for instance, as they can keep the memory of sequences of events. Recently, graph neural networks (GNN) have been successfully used for problems which can be framed in a graph structure.

Although the concept of neural network goes back to nearly three quarters of a century, their recent success (largely after 2012) is explained by the following trends (Fink et al. 2020):

- Advanced efficient algorithms, such as backpropagation. Many are now available freely on the Internet;
- The availability of huge amounts of data;
- The availability of affordable high-performance hardware (such as graphics processing units, GPU, for parallel computing) which makes it possible to process huge amounts of data very fast.

As a result, nowadays AI/machine learning and in particular deep learning, is permeating many fields of human activity. Particularly impressive have been the recent successes in image processing (the ImageNet, a database with more than one million labelled images), in machine translation, and, most recently, in Large Language Models (the ubiquitous Chat GPT and its competitors). The progresses in industry have been so far a little slower due to the particular challenges that need to be met (see e.g., Karim et al. 2023), but it is only a matter of time before industrial AI becomes widespread. In particular, since a survey paper on "potential and opportunities of deep learning for PHM" (Fink et al. 2020) appeared in 2020,

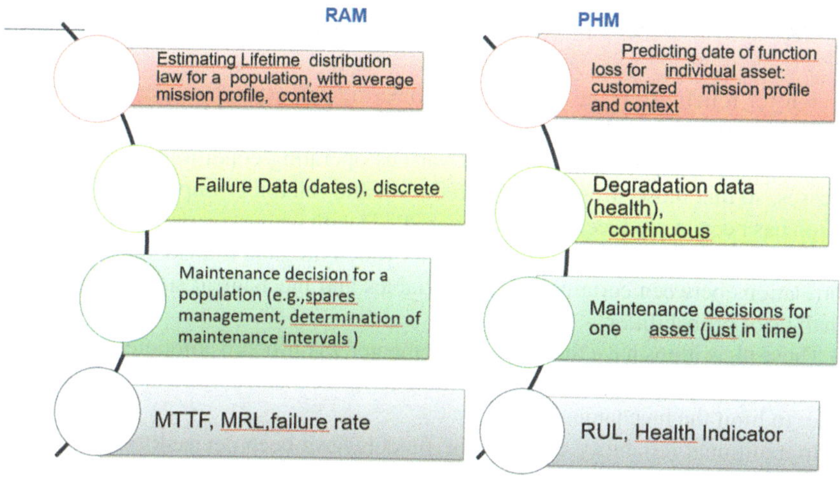

Fig. 4.3 Classical RAM (reliability-availability-maintainability) versus PHM (prognostics & health management)

applications have been multiplying in various industrial fields including railways, aerospace and electric power generation and transmission.

What are the implications for reliability engineering?

As a general statement, one could say that what AI enables, which traditional approaches could not, is the consideration of individual items beyond just population averages, and taking into account very precisely all the context variables that influence an item's degradation and failures. Figure 4.3 illustrates in that respect the main differences between classical RAM (Reliability-Availability-Maintainability) and PHM. Also, there is a key difference between traditional statistics and machine learning, as expressed by Breiman ("the two cultures", Breiman 2001): traditional statistics necessarily postulate an a priori probability model to describe the data, while machine learning explores the data without a priori.

The "ML revolution" offers tremendous opportunities in reliability engineering, which have only barely begun to be exploited. The field of reliability engineering (or more generally RAMS) and PHM (prognostics & health management) both are benefitting from ML and are actually coalescing into one single discipline.

The subject is vast and, in a lecture such as this, our modest goal is to give a few examples to try and illustrate the great potential of AI-ML as applied to reliability engineering and maintenance; while, at the same time, stressing the need for continuity and complementarity between traditional methods and AI-based ones. Necessarily, a number of important aspects had to be overlooked, and this is by no means an exhaustive survey of "AI for Reliability Engineering and Maintenance".

Design for Reliability and Reliability Prediction

Design for reliability is an important engineering activity, which can benefit from AI. For instance, in dealing with locomotives that operate in harsh conditions, it is important to understand the impact of various operating conditions on mechanical stresses, so that those stresses can be reduced, and reliability improved. This application has been treated recently (Gauthier 2022) with pattern recognition algorithms based on muti-layer feed-forward networks which are able to evidence very strong correlations between certain physical variables and mechanical stresses, with very small error probabilities.

Those algorithms have then been implemented in on-board computers, so that the axle force is adapted in real time according to the sensed operational conditions, in order to limit the mechanical stresses.

In traditional reliability engineering, models have been devised to represent the impact of various operational conditions, or stresses, on reliability. It is well known that environmental factors such as temperature or humidity impact reliability; but also, given a particular environment, asset mission profile will play a role as well.

Those factors, sometimes called covariates, have been incorporated in proportional hazard models, the best-known one being the Cox model (Cox 1972b):

$$\lambda(t; S) = \lambda_0(t)e^{\sum_{i=1}^{i=n} \beta_i S_i} \tag{4.1}$$

where λ denotes the failure rate (sometimes called hazard rate), S_i ($i = 1 \ldots. n$) denote the various stresses, and the coefficients β_i are to be estimated statistically.

In the Cox model (4.1), the failure rate is assumed to be proportional to a base failure rate $\lambda_0(t)$, and it further assumed that the coefficient of proportionality (i) is independent of time; (ii) depends linearly on the stresses. Those are of course simplifying assumptions, which are not necessarily verified in practice. For instance, they do not allow for modeling stresses that vary with time. Some more complex models have been introduced, whereby the coefficient can be a nonlinear function of the stresses, possibly time-dependent.

Recently, reliability researchers at Ford Motor Co. (Li et al. 2022) have introduced an AI-ML based method, inspired by machine translation. They have designed a special type of RNN, with an "attention mechanism"-the main idea behind which is to weigh all outputs of hidden states to dynamically highlight relevant features of the input data.

The goal is to exploit the ability of neural networks to learn highly complex, nonlinear functions. To do so, they adopt the viewpoint of translating time series, just as, in machine translation, a time series of written or spoken words in a given source language is translated into a time series of words in another, target language.

The source language here is the language describing asset status (i.e. the stresses, or observed features) at various points in time, and the target language is the language describing failure probability at various points in future, i.e. the reliability function, also called survival function (whose knowledge is equivalent to that of the failure rate). This is why the model is called a survival model, and it is a "deep survival" model because it relies on a deep (i.e. multi-layer) neural network.

This AI algorithm, call seq2surv2 (for "sequence-to-survival"), is able to make individual predictions on each asset based on that asset's individual exposure to stresses and operating conditions over time, which is much more powerful than the traditional reliability engineering approach which deals only with average population behaviors in static (i.e. non-time-varying) environments.

The learning scheme contains two aspects: feature extraction and survival function prediction. The architecture follows an encoder-decoder structure (see e.g., Doersch 2016); the encoder maps the input sequence to a latent state vector; the decoder generates the output sequence from the latent state vector. The method has been tested satisfactorily on the NASA C-MAPSS open data set (Arias et al. 2021), with excellent performance results. It must be emphasized though that the method is a "black box". At this stage, the predictions are not traced to identified failure modes or root cause analysis (such endeavors might be addressed in future).

Transfer Learning for Diagnostics and Prognostics

Data-driven machine learning models usually require (i) sufficiently many labeled data (so that the algorithms can be trained in a supervised way); (ii) identical distributions of data in the training set and the test set.

Particularly in industrial applications for diagnostics and prognostics, at least one of those two conditions is usually not fulfilled. The data on which an algorithm is trained often does not contain enough labelled data: concretely, a number of failure data have no identified root cause or failure code associated with them. Or, there are simply not enough failure data. Or the data on which the algorithms has been trained (the training set) are not really representative of the data to which it will be applied (the test set). Then, if some knowledge has been gained previously on systems that resemble the system under study, it is useful to transfer from those other systems whatever has been learned on them. This is the idea of transfer learning. For instance, one could transfer to one type of machine (e.g., an engine) what has been learned on a different machine (say, another engine type).Or, knowledge acquired in one context can be transferred to a different context: for instance, from the knowledge of a reliability function in accelerated stress conditions, derive the reliability function under standard, normal operating conditions. There are several methods for accomplishing transfer learning, an area of research which is progressing fast (see e.g., Yao et al. 2023) for an extensive survey).

In broad terms, they fall into the following three categories: (1) Model-based and parameter-based methods; (2) Feature-matching-based methods; (3) Adversarial adaptation methods.

(1) *Model-based and parameter-based methods*

Those methods take advantage of pre-trained models and adapt some of the model parameters to the new conditions. For instance, when the model is a deep neural network, weights from another application are used as initial weights, which cut the training time considerably. Thus the key idea is fine-tuning pre-trained parameters to the new application context.

(2) *Feature-matching-based Methods*

The key idea underlying those methods is to reduce the feature distribution difference across domains via feature transformation. The goal is to draw source-domain features and target-domain features closer to each other, so as to facilitate transfer (i.e. classification can be performed fairly easily in the target domain from features extracted from the data in the source domain).

There is a feature extraction step and a domain adaptation step.

(3) *Adversarial Adaptation Methods*

Those methods exploit a modified version of the General Adversarial Network (GAN) (Goodfellow et al. 2014). The generative network extracts features, and the discriminative network is used to tell the differences between source and target features. The goal is to learn features of one domain in such a way that the discriminator cannot distinguish them from features of the other domain. Domain-adversarial neural networks (DANN) belong to that category.

Industrial applications have been reported, for instance to turbines (see e.g., Michau and Fink 2019; Wang et al. 2019).

So far, there have been more applications to diagnostics than to prognostics. In any case, transfer learning has become a key enabler of advanced PHM systems and holds promises for reliability engineering (more generally RAMS) as well.

Combining Physics and Data

As illustrated in the previous sections, data-driven methods benefit from a number of recent breakthroughs and can be very efficient. However, as also mentioned, they suffer from a number of shortcomings. On the other hand, a number of physical degradation phenomena are well described by known physical laws, and it seems natural to use that knowledge when it exists, instead of just blindly processing data without paying attention to their meaning. Therefore, it seems that hybrid PHM, i.e., the joint use of physics knowledge with the processing of acquired field data, is a promising way forward. Physics-based and data-driven algorithms are complementary. A purely physics-based algorithm is limited by the need for a detailed knowledge of model

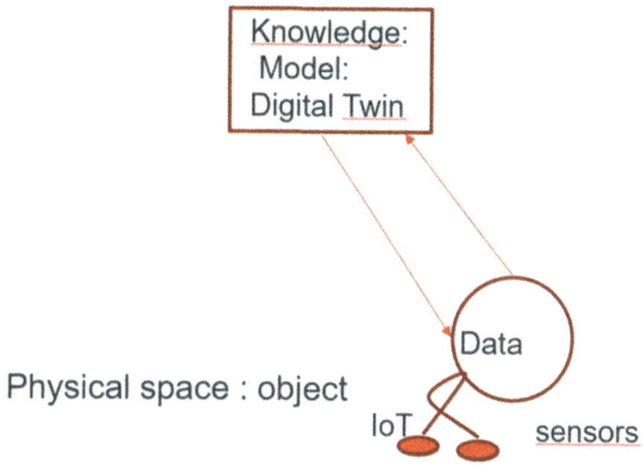

Fig. 4.4 Digital twin and physical space

parameters, some of which can vary over time with changing environmental conditions and evolving mission profile. In hybrid PHM systems, physical parameters are continually updated as new data is being acquired, as illustrated in Fig. 4.4.

The multi-physics model is sometimes called a digital twin.

A digital twin however is more than just a model. Just as any model, it is an abstraction of reality, i.e. it does not include all the details but only the parameters that are essential to the function being studied; at the same time, it ideally contains all relevant information relating to the history of the physical system; i.e., design changes, maintenance history, etc. A digital twin should accompany the physical system throughout its life time. Initially invented by NASA back in the Apollo program days, the concept has known considerable success recently, in PHM and RAMS applications in particular. Definitions vary and no unique standardized definition has emerged yet.

For instance, an example of definition is: "A Digital Twin is an integrated multi-physics, multi-scale simulation of a complex product which uses available models and information updates (such as sensor measurements, procurement and maintenance actions, configuration changes etc.) to mirror an asset during its entire lifecycle" (Karim et al. 2023).

IEEE distinguishes several classes of digital twins (IEEE 2020): the 'digital model', where changes in the physical object must be manually carried over to the digital twin; the 'digital shadow', where changes in the physical object are automatically carried over to the twin, and the full digital twin where changes occur automatically in both directions. Most digital twins in existence correspond to IEEE's 'digital shadow' concept.

A fairly recent illustrative example (Staino et al. 2018) is provided by the filter of an HVAC (heating-ventilation-air conditioning) system installed on tramway cars. HVAC performs a crucial function in warm climates. The filters tend to clog with dust accumulation, up to the point where the function of air exchange is no longer adequately performed. To avoid such a 'failure', filters are replaced preventively, and filter providers tend to be conservative in their recommendations for a replacement period. To avoid unnecessary replacements while avoiding loss of function, a tramway manufacturer (Alstom) has put in place a preventive maintenance strategy relying on the concept of digital twin. A physical law (Darcy's law) describes material accumulation. The model parameters are continually updated by means of pressure sensors: increasing clogging results in the need for a higher pressure differential upstream and downstream from the filter in order to achieve the same air flow through the filter. Continuous measurement of that pressure difference is the basis for the definition of a health indicator and the dynamic estimation of the filter's remaining useful life, i.e. the time left until reaching an unacceptable level of clogging.

This technique has led to a very accurate filter RUL prediction and it was evidenced that, under normal operating conditions, filter replacement periodicity could be halved without harm, with respect to supplier's recommendations.

That example is comparatively simple, in that there is a single failure mode (clogging) and the complete physical model is readily available (although not elementary at all) and lends itself well to the construction of a simple health indicator.

In general, a full physical model of all relevant degradations is not available.

Several approaches have been proposed to combine physics knowledge with data-driven methods (e.g., Arias-Chao et al. 2019).

The general idea is to use physics-based models to guide the discovery of useful machine-learning models, what is sometimes called "physics-informed machine learning" (Huber et al. 2023).

One promising approach (Arias et al. 2022) consists of estimating unobservable parameters from system dynamics (physics) and sensor readings. Those parameters encode the health condition of system components. They are then input into a deep neural network, along with sensor readings and physical model responses, to generate a prognostics model. This approach has been validated on a standard dataset, the 'Commercial Modular Aero-Propulsion System Simulation' (CMAPPS). It falls into the general category of "physics informed neural networks" (PINN), an active area of research.

A potential benefit of those hybrid approaches is to leverage the advantages of physics-based models and data-driven ones. Clearly, one should carefully avoid inheriting drawbacks from the two methods.

Quantifying and Managing Uncertainty

Reliability engineers have long been accustomed to dealing with uncertainty, typically by attaching a confidence interval to reliability, maintainability or availability estimates instead of just providing point estimates. In AI-based methods, especially when using neural networks, it is only recently that attention has been paid to uncertainty quantification. However, it is extremely important because several sources of uncertainty typically impact detection, diagnostics and prognostics metrics. Those include epistemic uncertainty, i.e., how much is unknown about the model (for instance the value of some model parameters), and what is sometimes called aleatory uncertainty, related to measurement errors and to variability in mission profile. Especially when estimating RUL (remaining useful life), which is affected among other by the future mission profile, providing confidence bounds is indispensable for risk management (Dersin 2023). In purely physics-based models, uncertainty quantification can be based on analytical methods such as first-order reliability methods (FORM) and first-order Taylor expansion of RUL based on the state equation (Sankararaman et al. 2014). For machine-learning based algorithms, more—complex methods have to be considered. A recent comprehensive tutorial (Nemani et al. 2023) surveys state-of-the-art methods, which include, among other, Gaussian process regression, Bayesian neural networks, and neural network ensembles. They lead to the notion of 'uncertainty-aware machine learning'.

Combining AI with Traditional Reliability Engineering

As already pointed out, we believe AI can enhance classical reliability engineering and that there should be a strong synergy between classical approaches and AI-based ones.

Let us illustrate this idea. Recently, this author introduced the use of a time transformation or time warping to describe the time evolution of equipment of system degradation (Dersin 2023). The time warping is in a one-to-one correspondence with the reliability function. It has the property that, in the transformed time, the mean residual life (MRL), i.e., the expectation of the RUL, is a linear function (Fig. 4.5).

The transformation (denoted g) also leaves invariant the first-and second-order moments of the time-to-failure distributions. In the transformed time, the slope of the MRL can be derived explicitly in terms of the TTF's coefficient of variation (i.e. the ratio of mean to standard deviation)-as in (4.2):

$$k = \frac{1 - \left(\frac{\sigma}{\mu}\right)^2}{1 + \left(\frac{\sigma}{\mu}\right)^2} \tag{4.2}$$

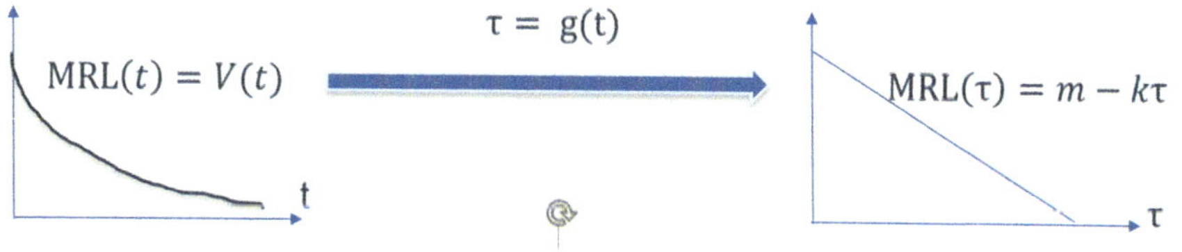

Fig. 4.5 Time transformation

where

$$\mu = \text{MTTF} \tag{4.3}$$

$$\sigma^2 = E\big[(TTF - \mu)^2\big] \tag{4.4}$$

That slope parameter k (a dimensionless quantity) is therefore an invariant of the time transformation g. It characterizes the speed of degradation somehow.

As the class of time-to-failure distributions with a MRL linear in time enjoys useful properties, among other an explicit formulation of the RUL confidence interval, those properties can be translated into equivalent properties for the initial distribution, using the inverse mapping g^{-1}. It is thus possible to derive explicit confidence intervals, and also bounds on the average time derivative of the RUL.

Now, identifying the time warping function can be performed by means of classical statistical methods (curve fitting) (Dersin 2023) but, if individual asset conditions, which typically evolve in time, must be taken into account, machine learning-including deep learning-algorithms are probably preferable because they make it possible to take into account a much larger number of parameters, dynamically.

Toward Optimized Dynamic Maintenance

The goal of predictive maintenance is to avoid failures as much as possible, while, at the same time, keeping the maintenance costs reasonable. In order to explicitly manage risks, one can impose an upper bound on the probability that RUL(t) is lower than the time to the next inspection. Let the next inspection occur at time $t + s*$. Then the constraint is

$$P\big[\text{RUL}(t) < s^*\big] < \alpha \tag{4.5}$$

In that way, s^* can be determined explicitly, as a function of t, using the time warping g(), by the method described in the previous section.

One can then select the value of the risk α in such a way as to minimize the sum of the total expected cost of maintenance (including preventive and corrective maintenance) and the expected cost of failures (Dersin 2023). More preventive maintenance will mean fewer failures, and therefore the optimal solution is a function of the relative costs (the cost of failure includes failures operational impacts, such as the costs of cancelled flights or of train delays, or lost production).

Since, in practice, environment and mission profile conditions, and therefore stresses, evolve with time, the estimates of the time warping g(.) and the 'degradation speed' k must be updated dynamically. They will also be influenced by the various maintenance operations that are carried out over time. Accordingly, a dynamic decision support process, described by the iterative procedure of Fig. 4.6, can be put in

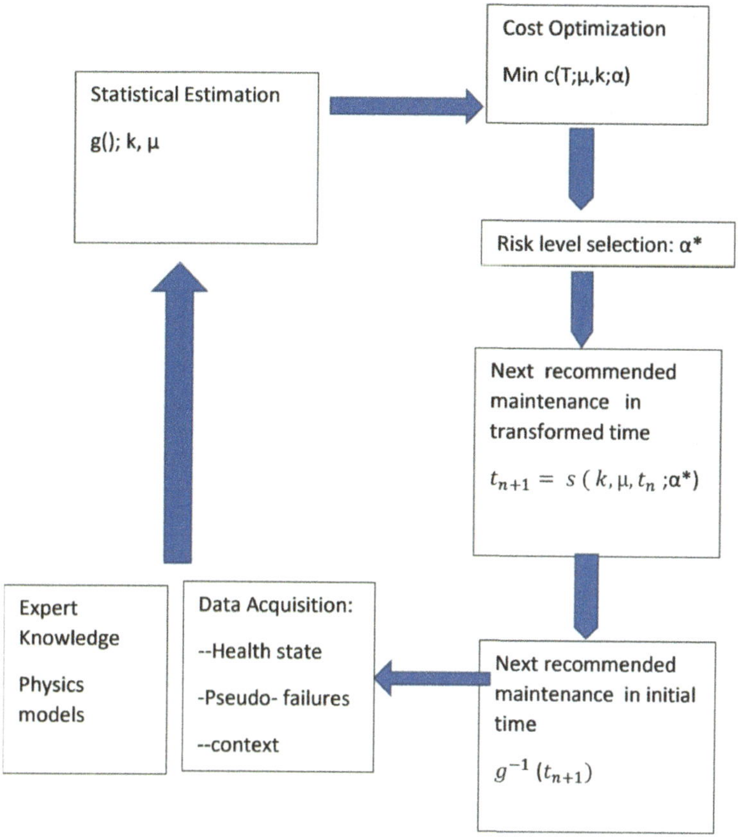

Fig. 4.6 Dynamic risk-based Predictive Maintenance. μ = MTTF. The cost function per unit of time is denoted c. (reproduced from (Dersin 2023) by kind permission of Taylor & Francis)

place. Identification of the g transformation and estimation of mean μ and variance (or mean and slope k) can also rely, not just on acquired field data, but on physical laws and expert knowledge as well. In our view at least, this process can only be a decision support tool and the ultimate decision maker is human.

Conclusions, Opportunities and Challenges

In this brief survey, it was only possible to scratch the surface of the burgeoning field of AI and its potential benefits for reliability engineering and maintenance management. For instance, reinforcement learning, natural language processing, large language models, are all about to revolutionize the field to some extent.

One of the challenges of AI in industrial applications (so-called 'Industrial AI') is interpretability, or explainability: how can a domain expert be convinced that the decisions recommended by a black box (typically a neural network) are justified? That would seem to require something like a 'white box' approach. Clearly, physics-based models are by definition more-easily explainable than purely data-driven ones. But as seen earlier, purely physics-based models are rarely available. The whole field of "explainable AI" (XAI), is therefore receiving increased attention (see (Arrieta et al. 2019) for a recent survey). For instance, SHAP (SHapley Additive exPlanations)—(Strumbelj et al. 2014; Ribeiro et al. 2016) is a game-theoretic approach to explain the output of any machine learning model, by determining the contributions of individual features on the algorithm's decisions. A methodology for focusing on causality rather than just correlation, named Causal Inference (Pearl 2009), is also part of that effort towards explainable AI.

In all algorithms, data quantity and quality is an important concern. Lack of data can sometimes be overcome by data augmentation, and plethora of data can be addressed by pre-processing to eliminate unnecessary or redundant data. Data quality management is a key activity in itself. Algorithms for tracking and eliminating corrupted or contaminated data do exist (see e.g., Ulmer et al. 2023). Another challenge, not the least one, is cybersecurity and data ownership. The more algorithms reside on the cloud, the more this subject comes to the fore (see e.g., Kour et al. 2022).

Industrial AI was the subject of a recent conference sponsored by Luleå University of Technology (IAI 2023), featuring among other the "AI Factory" (Karim 2022, 2023), a collaborative platform that allows multiple industrial partners to share only the data they need to share—in particular by bringing models to data rather than the opposite. That approach has shown its merit in railway applications, and is being generalized to other fields. Last but not least, in this era of climate change and emphasis on sustainable development, a key challenge is the energy consumption of algorithms-frugal AI is the corresponding 'buzzword". The notion combines that of energy frugality and low data consumption.

A good overview of PHM challenges can be found in (Zio 2022). A comprehensive treatment of state-of-the-art Reliability Engineering techniques can be found in (Birolini 2017) and (Nachlas 2017).

In conclusion, in spite of the very real challenges that still exist, especially in industrial applications of AI, we are of the opinion that reliability and maintenance engineers stand to benefit enormously from the potential of AI; and that, at the same time, it would be a mistake to believe that reliability engineering will "dissolve into AI".

References

Alpaydin E (2014) Introduction to machine learning. MIT Press, Cambridge, MA

Arias-Chao M, Adey D, Fink O (2019) Knowledge-induced learning with adaptive sampling variational autoencoders for open set fault diagnostics. arXiv:1912.12502v1

Arias Chao M, Kulkarni C, Goebel K, Fink O (2021) Aircraft engine run-to-failure dataset under real flight conditions for prognostics and diagnostics. Data 6(1):5

Arias Chao M, Kulkarni C, Goebel K, Fink O (2022) Fusing physics-based and deep learning models for prognostics. Reliabil Eng Syst Saf 217:107961

Arrieta et al (2019) Explainable artificial intelligence. arXiv:1910.10045

Barlow RE, Proschan F (1965) Mathematical theory of reliability. Wiley, New York

Birolini A (2017) Reliability engineering: theory & practice, 8th edn. Springer

Breiman L (2001) Statistical modeling; the two cultures (with comments and a rejoinder by the author). Stat Sci 16(3):199–231

Cox DR (1972a) The analysis of multivariate binary data. Appl Stat 113–120

Cox DR (1972b) Regression models and life-tables (with discussion). J R Stat Soc Ser B 34:187–202

Dersin P (2023) Modeling remaining useful life dynamics in reliability engineering. CRC Press, Taylor & Francis

(2020) Digital transformation. White Paper of the IEEE Digital Reality Initiative. DigitalReality@ieee.org

Doersch C (2016) Tutorial on variational auto-encoders. arXiv:1606.05908v2

Fink O, Wang Q, Svensén M, Dersin P, Lee W-J, Ducoffe M (2020) Potential, challenges and future directions for deep learning in prognostic and health management applications. In: Engineering applications of artificial intelligence, vol 92

Fisher RA (1922) On the mathematical foundations of theoretical statistics. Philos Trans R Soc Lond Ser A 222:594–604

Gauthier S (2022) Concrete applications of machine learning in railways. In: Proceedings of the ESREL 2022

Goodfellow IJ et al (2014) Generative adversarial nets. In: Proceedings of the international conference on neural information processing systems (NIPS 2014), pp 2672–2680

Hornik J, Stinchcombe M, White H (1989) Multi-layer feed-forward networks are universal approximators. Neural Netw 2:359–366

Huber LG, Palmé T, Chao MA (2023)Physics-informed machine learning for predictive maintenance: applied use-cases. In: 2023 10th IEEE Swiss conference on data science (SDS), Zurich, Switzerland, pp 66–72. https://doi.org/10.1109/SDS57534.2023.00016

Karim R, Galar D, Kumar U (2023) AI factory: theory, applications, case studies. Taylor & Francis, CRC Press

Karim A, Dersin P, Galar D, Kumar U, Jarl H (2022) AI factory: a framework for digital asset management. In: Proceedings of the ESREL 2022

Kour R, Patwardhan A, Thaduri A, Karim R (2022) A review of cybersecurity in railways. Proc Inst Mech Eng Part F: J Rail Rapid Transit

Li X, Krivtsov V, Arora K (2022) Attention-based deep survival model for time-series data. Reliabil Syst Saf 217

Michau G, Fink O (2019) Domain adaptation for one-class classification: monitoring the health of critical systems under limited information. arXiv:1907.09204v2

Nachlas J (2017) Reliability engineering-probabilistic models and maintenance methods, 2nd edn. Taylor & Francis, CRC Press

Nemani V et al (2023) Uncertainty quantification in machine learning for engineering design and health prognostics: a tutorial. arXiv:2305.0493

Pearl J (2009) Causal inference in statistics: an overview. Stat Surv 3:96–146. https://doi.org/10.1214/09-SS05

Ribeiro M, Sameer S, Carlos Guestrin C (2016) *LIME*: "Why should I trust you?: Explaining the predictions of any classifier. In: Proceedings of the 22nd ACM SIGKDD international conference on knowledge discovery and data mining. ACM

Sankararaman S, Daigle MJ, Goebel K (2014) Uncertainty quantification in remaining useful life prediction using first-order reliability methods. IEEE Trans Reliab 63(2):603–619

Staino A, Abou-Eïd R, Dersin P (2018) A Monte-Carlo approach for prognostics of clogging process in HVAC filters using a hybrid strategy—A real case study. In: Proceedings of the IEEE Conference on PHM
Strumbelj E, Igor Kononenko I (2014) Shapley sampling values. Explaining prediction models and individual predictions with feature contributions. Knowl Inf Syst 41(3):647–665
Ulmer M, Zgraggen J, Goren-Huber L (2023) A generic fully unsupervised framework for machine-learning-based anomaly detection. In: Proceedings of the ESREL 2023
Wang Q, Michau G, Fink O (2019) Domain-adaptive Transfer learning for fault diagnostics. arXiv: 1905.06004v1
Weibull W (1939) A statistical theory of the strength of materials. In: Proceedings of the Swedish Royal Institute of Engineering Research, p 153
Wiener N (1964) Extrapolation, interpolation, and smoothing of stationary time series: with engineering applications. MIT Press
Yao S et al (2023) A survey of transfer learning for machinery diagnostics and prognostics. Artif Intell Rev 56:2871–2922
Zio E (2022) Prognostics and health management (PHM): where are we and where do we (need to) go in theory and practice. Reliabil Eng Syst Saf 218(Part A)

Chapter 5
MIRCE Science: Solar Storm as a Mechanism of Motion of Autonomously Working Systems Through MIRCE Space

Jezdimir Knezevic

Introduction

Autonomously working systems can be defined as those that deliver at least one measurable function independently of human interaction by receiving its inputs from a set of electronic senses that are processed in accordance to establish algorithms. Thus, the ability to continuously communicate information required for safe operation is of vital importance to their operators.

On 27 February 2023 a powerful solar storm, containing a large amount of charged solar particles, reached Earth and SpaceX promptly delayed the planned launch of Starlink launch rocket. It was a direct result from the costly lesson learned from the loss of a batch of 40 satellites on 7 February 2022, which were launched into a moderately strong geomagnetic storm. As the loss of the Starlink satellites cost the company millions of dollars, SpaceX has not only started to pay greater attention to space weather forecasts since then, but also has been providing data from Starlink's onboard sensors to the U.S. National Oceanic and Atmospheric Administration (NOAA) to help them to improve its space weather forecasting models (Pultarova 2023).

The same solar storm of February 2023 temporarily disrupted operations of several Canadian oilrigs, which was the very first time in living memory that the drilling operations were suspended. The cause of the interruption was the impact of a solar storm on the accuracy of GPS signals that were driving the electronics in the part of the equipment which determines the direction and inclination the drill bit is going. The receiving information had unacceptable interference from the storm that those readings were unreliable (Pultarova 2023).

J. Knezevic (✉)
MIRCE Akademy, Exeter EX5 1JJ, UK
e-mail: jk@mirceakademy.com

P. V. Varde et al. (eds.), *Advances in Risk-Informed Technologies*, Risk, Reliability and Safety Engineering, https://doi.org/10.1007/978-981-99-9122-8_5

Events like this can occur up to 200 times during the eleven-year solar cycle and in the past have caused numerous problems to aircraft, power grid, spacecraft, communication and other working systems (Knezevic and Papic 2015; Zaczyk and Knezevic 2013; Murray et al. 2017). At the same time solar storms on the Sun are natural phenomena that have been happening for billions of years. Although they do not have dangerous impact to humans on Earth's surface, they could significantly reduce the reliability and safety of human designed and operated systems, failures of which could have catastrophic consequences on nature and humans.

Recent development of digital technologies enabled immense amounts of information to be compressed on small storage devices that can be easily preserved and transported. They have made fundamental changes to many aspects of human lives, including the creation of autonomously working systems. Autonomous ships, trains, cars and similar systems operate independently of human interactions, by receiving inputs information from the range of physical sensors that are processed in accordance to establish algorithms. Most frequently used sensors to control autonomous functions, include: global positioning system, inertial navigation system, optical and infra-red, light detection and ranging, radio detecting and ranging, microphones, including wind and pressure sensors. As the ability to continuously exchange information is essential for their functionality, these sensors are an integral part of a working system, as considered by MIRCE Science. The research conducted and published have shown that space weather, in general, and solar storm, in particular, have been impacting reliability and safety of a large number of modern technological systems, like power networks, aviation, satellite services, radio communication and pipelines, as documented in Knezevic and Papic (2015).

Consequently, the main objective of this paper is to draw attention to the MIRCE Science (Knezevic 2017) approach to reliability and safety of the autonomously working systems that could be beneficial to their designers regarding the potential impacts of the solar storms on the technology used. Then, and only then, meaningful reliability and safety engineering actions could be taken towards reduction of the probability of occurrences of damaging effects of solar storm on autonomous working systems during their in-service lives.

A Few Words About MIRCE Science

The philosophy of MIRCE Science is based on the premise that the purpose of the existence of any working system is to do expected work through time. The work is considered to be done when a measurable function(s) is performed. MIRCE Science focuses on the scientific understanding and description of the natural and human actions that govern the motion of working systems through MIRCE Space (Knezevic 2017). It is a conceptual 3-dimensional coordinate system containing a physical sequence of the motion of a given working system type through functionability states in respect to time, mathematically defined by corresponding convolution functions and corresponding probabilities. A full understanding of the mechanisms that generate this motion is essential for the accurate predictions of the expected performance of a given working system type using the mathematical scheme of MIRCE Science (Knezevic 2017).

According to MIRCE Science, at any instant of calendar time, a given working system could be in one of the following two macro states:

- Positive Functionability State (PFS), a generic name for a state in which a working system is able to deliver the expected measurable function(s),
- Negative Functionability State (NFS), a generic name for a state in which a working system is unable to deliver the expected measurable function(s), resulting from any reason whatsoever, including solar storms.

In MIRCE Science, work done by a working system is uniquely defined by the trajectory generated by its motion through MIRCE Space. That motion is driven by occurrences of functionability events generated by the following functionability actions:

- Negative Functionability Action (NFA), is a generic name for any natural process or human activity that compels a system to move to a NFS. Typical examples are: thermal ageing, actinic degradation, acid reaction, bird strike, warping, abrasive wear, suncups formation on the blue ice runway, fatigue, pitting, thermal buckling, photo-oxidation, production errors, strong wind, maintenance error, hail damage, lightning strike, COVID-19, quality problems in production or installation, tsunami, sand storm and so forth.
- Positive Functionability Action (PFA) is a generic name for any natural process or human activity that compels a system to move to a PFS. Typical examples are: servicing, lubrication, visual inspection, repair, replacement, final repair, examination, partial restoration, inspection, change of operational mode, modification, transportation, sparing, cannibalisation, refurbishment, health monitoring, packaging, diagnostics and similar

To scientifically understand the mechanisms that generate functionability actions, positive and negative, analysis of the in-service behaviour of several thousands of components, modules and assemblies of working systems in defence, aerospace, nuclear, transportation, motorsport and communication industries have been conducted at the MIRCE Akademy.

In MIRCE Science all negative functionability actions regarding individual components are categorised as following (Knezevic 2017):

- Component-internal actions (CIA) that consist of:
  - Inherent discrete actions, CIIA,
  - Cumulative continuous actions, CICA,
- Component-external actions (CEA), which are originated by:
  - Environmental phenomena, CEEA
  - Human activities, CEHA,

Corresponding negative functionability actions regarding the whole system are categorised as following (Knezevic 2017):

- System-internal actions (SIA)
- System-external actions (SEA):
  - Environmental phenomena, SEEA
  - Human activities, SEHA

Research studies conducted by the author had shown that any serious studies of the functionability mechanisms have to be based between the following two boundaries (Knezevic 2017):

- the "bottom end" of the physical world, which is at the level of the atoms and molecules that exists in the region of 10^{-10} of a metre.
- the "top end" of the physical world, which is at the level of the solar system that stretches in the physical scale around 10^{+10} of a metre.

This range is the minimum sufficient "physical scale" which enables scientific understanding of relationships between physical phenomena that take place in the natural environment and the physical mechanisms that govern functionability events during the life of working systems.

In summary, MIRCE Science is a theory of the motion of working systems through MIRCE Space resulting from any functionability actions whatsoever and the actions required to produce any functionability related motions. To that aid a scientific understanding of mechanisms that generate positive and negative functionability events is an imperative. Without full understanding of these mechanisms the prediction of occurrences of functionability events is not possible, and without the ability to predict the future, the use of the word science becomes inappropriate.

Solar Storm as a Negative Functionability Actions

The Sun is a magnetic variable star at the centre of the Solar system that drives the space environment of the planets. It continuously undergoes changes, some of which could be extremely violent. These changes are transferred by the solar wind to the Earth, where they disturb its magnetic field. The regular changes in the level of solar activity over long-periods are known as the solar cycle. These cycles vary between 9.5 and 11 years. Usually, the number of sunspots on the solar surface measures solar activity. The solar cycle is also seen in the number and strength of the solar flares, which are resulting from tremendous explosions in a localised region on the Sun. In a matter of just a few minutes they heat material to many millions of degrees (Pultarova 2023).

Solar Storms happen when a Sun emits large bursts of energy in the form of solar flares and coronal mass ejections. These phenomena send a stream of electrical charges and magnetic fields towards the Earth at high speed, in the form of x-rays, ultraviolet and radio emissions, which can cause disruptions to the Earth's ionosphere leading to radio and communications interference.

One of the effects of a solar storm striking Earth is the creation of the "northern lights" which are seen in the regions around the Arctic Circle. An adverse effect of solar storms is the disruption of satellites and other electronic means of communications (Wernik n.d).

Types of Solar Storms

Solar Storms come in the form of the following types (Dobrijevic 2022):

- Solar Flares, which are manifested as a sudden flash of increased brightness on the Sun, usually observed near its surface and in proximity to a sunspot group. Powerful flares are often, but not always, accompanied by a coronal mass ejection. Even the most powerful flares are barely detectable in the total solar irradiance.
- Coronal Mass Ejections (CME), which are a result of the twisting and realignment of the sun's magnetic field. As magnetic field lines "tangle" they produce strong localised magnetic fields that can break through the surface of the Sun at active regions. It is manifested through a significant release of plasma and accompanying magnetic field from the solar corona. They often follow solar flares and are normally present during a solar prominence eruption.
- Geomagnetic Storm, which is a temporary disturbance of the Earth's magnetosphere caused by a solar wind shock wave and/or cloud of magnetic field that interacts with the Earth's magnetic field, caused by changes in the solar wind and interplanetary magnetic field (IMF) structure.
- Solar Particle Events, or solar proton event (SPE), occurs when particles (mostly protons) emitted by the Sun become accelerated either close to it during a flare or in interplanetary space by coronal mass ejection shocks.

Coronal Mass Ejections on the Sun

Coronal mass ejections are large bubbles of coronal plasma threaded by intense magnetic field lines that are ejected from the Sun over the course of several hours. CMEs often look like huge, twisted rope, which is commonly known as "flux rope" (Freeman 2001).

CMEs disrupt the flow of the solar wind and cause disturbances that can damage systems in near-Earth and on Earth's surface. Their magnetic fields merge between the interplanetary magnetic field (IMF) and geomagnetic field lines (NASA Space Technology n.d). This direct link between even a small percentage of the geomagnetic field lines and the IMF results in large increases in the rate of energy transfer from the solar wind and the magnetosphere. Consequently, CMEs are among the most important drivers of geomagnetic storms and sub-storms. For example, in February 2011, a CME produced by an especially-powerful solar flare disrupted radio communications throughout China. There estimates that a major solar storm could cause up to twenty times more economic damage than the worst hurricane.

Impacts of Solar Storms on Earth

Solar Storms can have significant impacts on Earth that could be summarised as following:

- radiation poisoning to humans and other mammals caused by very high-energy particles, such as those carried by CME
- temporary disturbance of the Earth's magnetic field caused by CME strikes on Earth's atmosphere. It can throw satellites off course and cause them to fall to the surface of the earth, putting many urban centres at risk.
- impact on migrating animals that use magneto reception to navigate, such as birds and honey bees,
- induction of currents in pipelines by rapidly fluctuating geomagnetic fields that can cause the flow meters to transmit erroneous flow information, on one hand, and dramatic increase of the corrosion rate, on the other.

Impacts of Solar Storms on Sensors that Control Autonomous Functions

Recent developments of digital technologies have made fundamental changes to the way we live our lives. Among others, digital technology is a driving force behind all autonomously working systems, from passenger cars to spacecraft. In general, autonomously working systems can be defined as a set of entities that can operate independently of human interactions. They are performing expected functions by receiving inputs from a set of electronic senses that are processed in accordance to establish algorithms. Thus, the ability to continuously exchange information relevant to the safe operation of autonomously working systems is of vital importance for their functionability (Knezevic and Papic 2015).

In the context of MIRCE Science an autonomously working system contains, among many other entities, a communication and control parts that provide autonomy.

According to Kim et al. (2020) the following elements are driven by a digital technology similar to other self-driving means of transport, which use a range of physical sensors to control autonomous functions, such as:

- Global Positioning System (GPS)
- LIght Detection And Ranging (LIDAR)
- optical and Infra-Red (IR) cameras;
- Inertial Navigation System (INS)
- RAdio Detecting And Ranging (RADAR)

Consequently, the functionability of autonomously working systems depends also on the reliability and accuracy of the controlling sensors listed above, which in turn, depend on reliability of satellites involved and technology used.

Solar Storm as a Negative Functionability Event in MIRCE Science

Negative functionability action is a generic name for any natural process or human action that compels a working system to move to a NFS, some of which are mentioned earlier in the paper. The main objective of this paper is to introduce solar storm as one of the numerous natural actions that have a negative impact on the functionability of an autonomously working system. Their physical occurrences are manifested through occurrences of negative functionability events. These are observable instances of time at which working systems sees to deliver expected measurable function(s) (Zaczyk and Knezevic 2013).

When energetic protons, generated by a solar radiation storm, collide with autonomously working systems in space, they can penetrate deep into them and could generate NFE in their electronic circuits. Also, when the energetic protons collide with the atmosphere, they ionise the atoms and molecules thus creating free electrons. These electrons create a layer near the bottom of the ionosphere that can absorb high frequency (HF) radio waves making radio communication difficult or impossible.

Frequencies of Solar Storms

Regarding the frequencies of occurrences geomagnetic storms, generated by CME, today the following two classes are distinguishable:

- "recurrent", corresponding with the Sun's rotation, occurring every 27 days.
- "non-recurrent", whose frequency is variable through time

As earlier stated, the frequency of occurrence of solar flares varies with the solar cycle, which can range from several per day during solar maximum to less than one every week during solar minimum. More powerful flares are less frequent than weaker ones. The probability of an extreme space weather event having a direct impact on Earth is between 0.016 and 0.12 every decade.

Placing Solar Storm in MIRCE Functionability Equation

MIRCE Functionability Equation is a mathematical description of the motion of the motion of working systems through MIRCE Space, defined by Knezevic (2014), thus:

$$y(t) = 1 - \sum_{i=1}^{\infty} F_S^i(t) + \sum_{i=1}^{\infty} O_S^i(t) \tag{5.1}$$

In the above equation $F_S^i(t)$ is a cumulative distribution function of the random variable that mathematically represents the time to the occurrence of the ith negative functionability event, $TNE_S^i(t)$ of a system considered. In MIRCE Science it is defined by a following convolution integral:

$$F_S^i(t) = \int_0^t O_S^{i-1}(x) dF_{s,i}(t-x), \quad i = 1, \infty \tag{5.2}$$

where $F_{S,i}(t)$ is a cumulative distribution function of the random variable that mathematically represents the time to the occurrence of the ith negative functionability event, $TNE_{S,i}(t)$ of a system or component considered. In the case that this random variable is governed by the impact of a solar storm on the autonomously working system, it is denoted as $TNE_{S,i,SS}$, and it is defined by the following expression:

$$F_{S,i}(t) = P(TNE_{S,i,SS} \leq t) = \int_0^t f_{S,i,SS}(t) dt \tag{5.3}$$

where $f_{S,i,ss}(t)$ is a probability density function of the random variable that defines the time to the occurrence of ith negative functionability event, which in this specific example is solar storm. The above equation is in the most generic form and as such covers all possible variations and behaviours of solar storms, which means that its users have to determine the applicable mathematical expressions for their specific application.

In the Eq. (5.1) $O_S^i(t)$ is a convoluted form of cumulative distribution function of the random variable that mathematically represents the time to the occurrence of the ith positive functionability event, $TPE_S^i(t)$ of a system or component considered. In MIRCE Science it is defined by the following convolution integral

$$O_S^i(t) = \int_0^t F_S^i(x) dO_{S,i}(t-x), i = 1, \infty \quad (5.4)$$

where $O_{S,i}(t)$ is a cumulative distribution function of the random variable that mathematically represents the time to the occurrence of the ith positive functionability event, $TPE_{S,i}(t)$ of a system or component considered. In the case that this random variable is governed by the impact of a recovery action from occurred solar storm on the autonomously working system. It is denoted as $TPE_{S,i,SS}$, and it is defined by the following expression:

$$O_{S,i}(t) = P\left(TPE_{S,i,SS} \le t\right) = \int_0^t o_{S,i,ss}(t) dt \quad (5.5)$$

where $o_{S,i,ss}(t)$ is a probability density function of the random variable that defines the time to the occurrence of ith positive functionability event, which in this specific example is solar storm recovery action. The above equation is in the most generic form and as such covers all possible positive functionability actions that could be taken to return a system to PFS, after solar storms.

In summary, it is essential to stress the following two points:

- the above presented equations are a generic mathematical interpretation of the physical reality of the functionability of working systems. However, the accuracy of their predictions are in the hands of their users, whose knowledge and understanding of the physical reality guide them to the selection of the most appropriate mathematical functions to represent the impacting natural and human actions
- the impact of solar storm could be at:
  - the component level, CEEA, that might or might not affect the functionability of a working system considered
  - the system level, SEEA, that will affect the functionability of a system without a failure of any individual consisting component, for example, the cases of disappearances of the whole satellites.

Solar Storm as Mechanism of Motion of Satellites Through MIRCE Space

Satellites are critically important for the successful operation of the most of the modern technological world to function as expected. Thus, it is essential to understand all different mechanisms that solar storms and consequential geomagnetic storms threaten orbiting satellites that are essential for reliable and safe operation of autonomously working systems, among others.

When the Earth atmosphere absorbs energy from magnetic storms, it heats up and expands upward. This expansion significantly increases the density of the thermosphere, the layer of the atmosphere that extends from about 80 km to roughly 1000 km above the surface of Earth. Higher density means more drag, which could cause a problem for the motion of satellites increasing the probability of the transition to NFS of the working system affected.

Drag is just one hazard that space weather poses to space-based technological systems. Strong geomagnetic storms generate the significant increase in high-energy electrons within the magnetosphere, which that penetrate the shielding on a spacecraft and accumulate within its electronics. The build-up of electrons can discharge in small lightning strikes and damage electronics, generating occurrence of NFE.

Penetrating radiation or charged particles in the magnetosphere, even during mild geomagnetic storms, can also alter the output signal from electronic devices. This phenomenon can cause errors in any part of a spacecraft's electronics system, and if the error occurs in critical parts, the entire satellite can experience that final transition to NFS. It is necessary to stress that small errors are common and usually recoverable by corresponding PFA, although terminal failures, though rare, do happen.

Finally, a solar storm can disrupt the ability of satellites to communicate with Earth using radio waves. Many communications technologies, like GPS, for example, rely on radio waves. As the atmosphere continuously distorts radio waves by some amount, design engineers correct for this distortion when building communication systems. However, during geomagnetic storms, changes in the ionosphere, the charged equivalent of the thermosphere that spans roughly the same altitude range, will change how radio waves travel through it. The calibrations in place for a quiet atmosphere become inadequate during solar storms.

First Solar Storm Simultaneously Impacts Earth, Moon and Mars

A coronal mass ejection that took place in August 2021 sent simultaneously energetic particles to Mars, the Earth and the Moon, emphasising the need to prepare human space missions for the dangers of solar radiation (Lea 2023). This CME caused an influx of highly energetic, and thus fast-moving, charged particles across the surface of these solar system bodies.

The detection of the same coronal mass ejection on these three different worlds for the first time highlighted the necessity for a better understanding of a mechanism that drives interaction between a planet's magnetic field and atmosphere. This is instrumental for the determination of methods for shielding autonomously working systems from such radiation.

The 2021 CME detection was the very first-of-its-kind. Interestingly, at the moment of the eruption, Earth and the Red Planet were on opposite sides of the Sun with a distance between them of around 250 million km. This outburst was detected by the:

- Euglena and Combined Regenerative Organic-Food Production in Space (Eu:CROPIS) orbiter around Earth.
- ExoMars Trace Gas Orbiter (TGO) on Mars,
- Chang'e-4 Moon lander and NASA's Lunar Reconnaissance Orbiter (LRO) on the lunar surface

In October of the same year a coronal mass ejection was observed by the ESA/NASA **SOHO** observatory represented by a rare event called a "ground level enhancement" during which charged particles from the Sun travel fast enough to penetrate the magnetosphere and reach the ground. This particular occurrence is just the 73rd example of such an event since records began in the 1940s, and it remains the last to be recorded (Space and exploration n.d).

As Mars and the Moon do not have a magnetic field, charged solar particles can strike their surfaces more often than on Earth generating a secondary radiation from their surfaces. However, the atmosphere on Mars, while much thinner than Earth's, can still stop low-energy particles and slow high-energy particles.

With both Mars and the Moon being the next human space exploration for future crewed space exploration, it is essential to know how both locations are impacted by solar radiation and, thus predict what humans might experience on the surfaces of those worlds if/when astronauts embark on long-term missions there in the future.

In August 1972, a solar outburst occurred that would have delivered precisely such a high dose of radiation to an astronaut if one had been on the moon at the time, but fortunately, it happened between the crewed Apollo's missions 16 and 17.

Protection from Space Storms

Digital technology is critically important for much of the modern world to function, and protecting space assets from space weather and solar storms is an important area of concern. Some of the risks can be minimised by shielding electronics from radiation or developing materials that are more resistant to radiation. According to Murchison (2023) the most frequently used materials for electromagnetic shielding include: sheet metal, metal screen, and metal foam. Any holes in the shield or mesh must be significantly smaller than the wavelength of the radiation that is being kept out, or the enclosure will not effectively approximate an unbroken conducting surface. However, there is only so much shielding that can be done in the face of a powerful geomagnetic storm (Murray et al. 2017).

The ability to accurately forecast storms would make it possible to pre-emptively safeguard satellites and other assets to a certain extent by shutting down sensitive electronics or reorienting the satellites to be better protected. While the modelling and forecasting of geomagnetic storms has significantly improved over the past few years, some of the projections have not been confirmed in physical reality.

The NOAA had warned that, following a coronal mass ejection, a geomagnetic storm was "likely" to occur the day before or the day of the February 2022 Starlink launch, but the mission went ahead anyway.

Conclusions

Autonomously working systems are rapidly expanding. For example, maritime systems have been exposed to autonomy from surface to underwater vehicles being deployed for patrol, oceanographic and maintenance among other purposes. Furthermore, cargo ships projects involving coastal and ocean-going routes with different degrees of autonomy are being tested. In October 2021, the International Maritime Organization (IMO) approved an output to develop regulation for Maritime Autonomous Surface Ships (MASS) (Miskovic et al. 2018).

This paper has shown that a solar storm has impacted reliability and safety of a large number of modern technological systems, through real life examples. Hence, this fact led the author to conclude that a solar storm could have similar impacts on the reliability and safety of future autonomously working systems.

Consequently, this paper briefly examined the solar storm phenomena that could generate undesirable consequences on the working life of autonomously working systems as guidance to their future designers and operators. Then and only then, accurate reliability and safety predictions could be possible, by applying methods of MIRCE Science, to achieve the ultimate goal of increasing the probability of the prevention and protection of occurrences of undesirable events generated by naturally occurring space storms phenomena on the autonomously working systems. The description of MIRCE Functionability equations is given with specific emphasis of including that impact of solar storm actions, at component and system level, for the prediction purposes.

Solar storms are natural phenomena that will continue to follow their own course. As they have evidential impacts on operation of autonomously working systems through functionality of digital technologies, it is essential to understand them better. Thus, the future research that should enable safer and more reliable operation of autonomously working systems could go in two directions. First, further improvement of the solar storms prediction services towards provisioning of the early warnings in time sufficient for the preventive actions to be taken by potentially endangered working systems. The second direction for the future research should be focused on innovative technologies and methods for designing equipment that is able to operate safely or protect autonomously working systems in the events of harmful impacts of solar storms in the future.

References

Dobrijevic D, Coronal mass ejections: what are they and how do they form? Space.com. https://www.space.com/coronal-mass-ejections-cme. 24 June 2022

Freeman JW (2001) Storms in space. Cambridge University Press, Cambridge, UK, pp 140

Kim M, Joung T, Jeong B, Park H (2020) Autonomous shipping and its impact on regulations, technologies, and industries. J Int Marit Saf Environ Aff Shipping 4(2):17–25. https://doi.org/10.1080/25725084.2020.177942

Knezevic J (2017) The origin of MIRCE science. MIRCE Science, Exeter, UK, pp 232, ISBN 978-1-904848-06-6

Knezevic J (2014) MIRCE functionability equation. Int J Eng Res Appl 4(8) (Version 7):93–100. www.ijera.com. ISSN 2248-9622

Knezevic J, Papic Lj (2015) Space weather as a mechanism of the motion of functionability through life of industrial systems. Adv Ind Eng Manage 4(1):1–8, American Scientific Publishers, Printed in the United States of America. Print ISSN 2222-7059; Online ISSN 2222-7067

Lea R (2023) https://www.space.com/expansive-solar-eruption-illustrates-risk-of-radiation-for-future-space-missions. Published 3 Aug 2023

Miskovic et al. (2018) Impact of technology on safety as viewed by ship operators. Trans Marit Sci 01:53–58. https://doi.org/10.7225/toms.v07.01.005

Murchison J (2023) Protecting your electronics from EMP and solar storms. In: Circuits electronics. https://www.instructables.com/Protecting-Your-Electronics-From-EMP-and-Solar-Sto/. Accessed 27 Aug 2023

Murray SA, Bingham S, Sharpe M, Jackson DR (2017) Flare forecasting at the Met Office Space Weather Operations Centre. Space Weather 15:577–588. https://doi.org/10.1002/2016SW001579

NASA Space Technology, https://www.jpl.nasa.gov/nmp/st5/SCIENCE/cme.html (downloaded 4.9.2023)

Pultarova T, Powerful solar storm delays SpaceX rocket launch, stalls oil rigs in Canada amid aurora-palooza, Space.com. https://www.space.com/solar-storm-delays-spacex-launch-supercharges-auroras. 28 Feb 2023

Science & Space exploration, https://www.esa.int/ESA_Multimedia/Images/2023/08/Coronal_mass_ejection_on_28_October_2021

Wernik A, What is space weather? space research centre. Polish Academy of Sciences. Warszawa, Poland, pp 27–32. http://www.cas.uio.no/Publications/.pdf

Zaczyk I, Knezevic J (2013) Cosmic phenomena in MIRCE mechanics approach to reliability and safety. SRESA's Int J Life Cycle Reliab Saf Eng 2(2): 41–50. Mumbai, India, ISSN 22500820

Chapter 6
The Development of the Integrated System Failure Analysis and Its Applications

Carol Smidts and Xiaoxu Diao

Introduction

This paper reviews the development of ISFA, i.e. the Integrated System Failure analysis and its related domain of application. Section 2 discusses the progressive development of the approach and its direct application to various types of systems of various sizes in terms of components and the number of faults being simulated. Section 3 will discuss the domain of relevance and potential applicability of ISFA. Section 4 will provide our conclusion.

Progressive Development of ISFA and Applications

The ISFA approach (Mutha et al. 2013) was proposed to enhance traditional fault analysis techniques for the early design stage of analysis of complex systems containing hardware, software, and their interactions. ISFA integrates two existing techniques—the Functional Failure Identification and Propagation (FFIP) approach for hardware systems and the Failure Propagation and Simulation Approach (FPSA) for software systems. The integrated model allows concurrent hardware and software simulation to identify failure propagation paths and their functional impacts. Component behavioral rules including nominal and faulty modes are defined. The function failure logic relates component behavior to the system function status. The key benefits of ISFA include early design stage integrated analysis, identifying cross-domain

C. Smidts (✉) · X. Diao
The Ohio State University, Columbus, USA
e-mail: smidts.1@osu.edu

X. Diao
e-mail: diao.38@osu.edu

P. V. Varde et al. (eds.), *Advances in Risk-Informed Technologies*, Risk, Reliability and Safety Engineering, https://doi.org/10.1007/978-981-99-9122-8_6

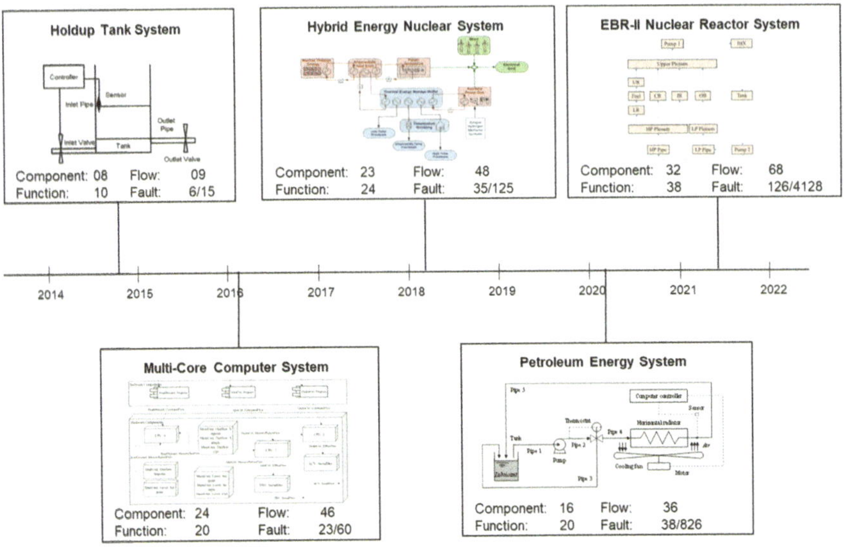

Fig. 6.1 Systems used in applications of ISFA and characteristics of the related ISFA models

failure propagation, handling multiple concurrent faults, and automating propagation analysis. A holdup tank system (see Fig. 6.1) was used initially as a case study system to verify its effectiveness.

As a first derivative application of ISFA, the method was applied as a fault propagation and effects analysis approach to help design an online monitoring (OLM) system for the secondary loop of a nuclear power plant in a hybrid energy system (Diao et al. 2018). The overarching goal of the research was to analyze potential faults in critical components of the system and how their related disturbances would propagate through the system to determine the optimal location in which to position sensors and determine which specific signals should be measured for effective fault detection and diagnosis. The analysis used ISFA to construct models and simulate fault propagation through qualitative physics-based rules (Forbus 1988). The models developed include configuration models depicting hardware/software components and functional models showing flows between components to achieve system goals. Component behavioral rules define nominal and faulty input–output relations. Functional failure logic assesses component states during simulation. The proposed approach was applied more specifically to a thermal energy storage subsystem of hybrid energy nuclear systems. Fault simulations injected single and multiple faults to identify new system behaviors and evaluate signals for distinguishing fault scenarios. Experiments on a hardware-in-the-loop system were used to validate the simulation results. The method provides valuable information early in the design process when detailed data is lacking, including assessing signal diagnosability for faults and guiding sensor selection to optimize the OLM system. Enhancements like adding precise parameter ranges can improve the accuracy of the approach and related results over time as the design of the system matures. Overall, the physics-based simulation approach helps design effective monitoring systems for new complex energy systems by leveraging qualitative component knowledge during early development stages. The results can focus on sensor locations and enable fault diagnosis capabilities in the OLM system.

Next, we then developed a propagation-based fault detection and discrimination (PFDD) method for detecting and diagnosing faults based on simple features generated through the execution of ISFA models (Li et al. 2022). The PFDD method has several advantages over traditional model-based and data-driven fault diagnosis techniques. It utilizes more information than analytical redundancy relations (Chow and Willsky 1984), works for both steady state and transients, has a low computational cost, and enables sensor optimization early in the design process. The PFDD method and sensor optimization were applied to an ISFA model of the Experimental Breeder Reactor II (EBR-II) as a case study. The simulation results show that the method can discriminate four types of example faults using only a single optimally placed sensor despite the complexity of the system model. The case study demonstrates the potential to diagnose faults with fewer sensors than conventional techniques. The features used to characterize fault propagation include the deviation of variables from expected values, the variation of variables over time, and the order in which variables are influenced. These features are extracted by simulating the ISFA model under different fault conditions and are used to generate fault detection and diagnostic strategies.

In addition, ISFA has been integrated with an ontology-based approach leading to the Integrated System Failure Analysis using an ONtological framework (IS-FAON) which can be used to analyze computer systems' fault propagation and effects (Diao et al. 2022). The approach uses ontologies to model the system components, functions, flows, faults, and their relationships. It allows for *generating* and injecting diverse types of faults into the system model and inferring the propagation of the deviations induced by the faults and inferring the nature of their impacts on system functionality without detailed implementation information. Based on the information modeled by ISFA, the ontologies are used to define relevant concepts like components, flows, functions, states, behaviors, and faults. Dependency rules between concepts are specified. Principles are proposed to generate new types of faults never experienced before by modifying the properties of components. During the simulation, faults are injected into the system model by modifying component states and fault propagation paths can be inferred by executing component behaviors chronologically and deducing component/function states at each timestep.

As another application of ISFA, we developed an expert-based method called Fuzzy Functional Failure Identification and Propagation (Fuzzy-FFIP) to analyze functional failure risk in fracturing systems (Wang et al. 2021). The method integrates functional failure analysis (using a system model with configuration, behavior, and function) with fuzzy logic to handle imprecise relationships between component and functional states. Fuzzy logic is used to assess the fuzziness of functional states (operating, degraded, lost) based on deviations in key variables like flow rate and temperature. Fuzzy-FFIP is applied to analyze failure risk in the lubrication subsystem of a fracturing pump truck. Fuzzy membership functions define transitions between functional states based on variable deviations. Monte Carlo simulation allows risk analysis for single and double faults. Results show that Fuzzy-FFIP provides more refined risk levels compared to traditional functional failure analysis. It reduces uncertainty in the risk estimates.

Figure 6.1 displays details on the target systems of those ISFA applications and the characteristics of the created ISFA models (e.g., the numbers of components, flows, functions, and faults).

Based on ISFA and propositional logic, our more recent research has focused on a method for backward failure propagation to identify the possible causes of a known system failure (Mansoor et al. 2023). The method can formally derive reversed rules from the forward ISFA rules. These reversed rules allow propagating a known failure state backward through the system to deduce the component modes and functional dependencies that could have led to the failure. The backward propagation starts with one or more known functional failure states and traces back to find the combinations of component faults and physical variable relationships that could satisfy the initial failure conditions. At each time step, the analysis considers the physics-based behavioral rules and constraints to eliminate invalid combinations and propagates only the valid modes and physical variable relations backward. In comparison to other physics-based approaches, the key strengths of this formal deductive method are its mathematical provability using propositional logic, its higher abstraction level working backward from functional to component level, and its exhaustive deduction of all valid system states for the given failure condition. The analysis is automated using a Satisfiability Modulo Theories (SMT) solver. The method is broadly applicable across engineering domains like nuclear, aerospace, electronics, etc.

Figure 6.2 displays the development timeline of ISFA.

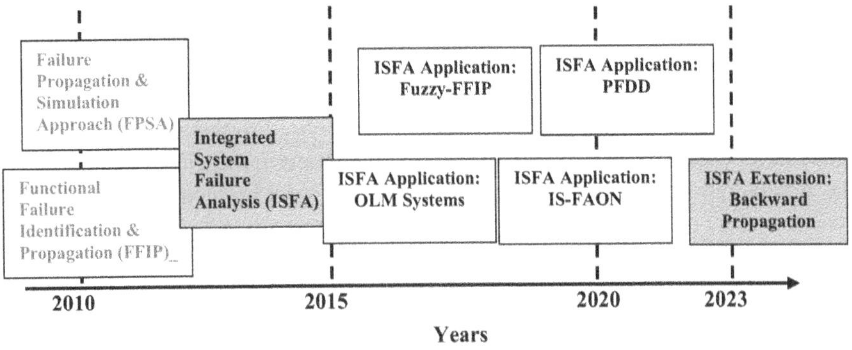

Fig. 6.2 Timeline of the development of ISFA

ISFA and Its Potential Domain of Applicability

To determine the potential domain of applicability of ISFA, we have referred ourselves to the research literature citing our work. Our assumption has been that since these works considered our research relevant to some degree to their area of study, this might be a possible direction in which to grow. The methodology followed to build the landscape of relevant areas of applicability used a combination of systematic literature review and qualitative analysis. More specifically, we extracted all relevant works that cited the six papers given in Table 6.1 which we consider to be foundational to our work on ISFA. These references include reports, thesis, conference papers and journal papers. These were extracted using Publish or Perish (Harzing 2007) from relevant databases, Google Scholar, Web of Science, and Scopus on 09/07/2023. This resulted in 104 references which are provided in the reference section of this paper as references (César Sobrinho et al. 2016) to Pietrykowski (2022). The extracted references contained 4 non English papers (César Sobrinho et al. 2016; 박진희, 백종문 et al. 2011; 王恩亮 and 张丽华, "嵌入式系统软件可靠性模型的研究" 2014; 陆寅, 秦树东, 郭鹏, and 董云卫, "软硬件综合 AADL 可靠性建模及分析方法 (形式化方法与应用)", 软件学报) which we eliminated from our pool. In addition, two papers (Alidousti et al. 2011; Shine et al. 2017) seemed to have been extracted in error. These were also removed from our pool.

This resulted in 98 papers. For each such papers, the titles of the papers were extracted and used in the qualitative analysis. The qualitative analysis followed the approach defined in Saldaña (2021). One of the authors performed a preliminary analysis the text (i.e. of the titles) which was reviewed by the second author. The analysis focused on the themes of 'purpose' and 'domain'. Other themes can potentially be further extracted from the titles but will not be discussed in this paper.

Table 6.1 Papers used as a basis for extraction of ISFA relevant documents

Ref. No.	Titles
Mutha et al. (2013)	An integrated multidomain functional failure and propagation analysis approach for safe system design
Diao et al. (2018)	Fault propagation and effects analysis for designing an online monitoring system for the secondary loop of the nuclear power plant portion of a hybrid energy system
Li et al. (2022)	A propagation-based fault detection and discrimination method and the optimization of sensor deployment
Diao et al. (2022)	An ontology-based fault generation and fault propagation analysis approach for safety–critical computer systems at the design stage
Wang et al. (2021)	An expert-based method for the risk analysis of functional failures in the fracturing system of unconventional natural gas
Mansoor et al. (2023)	A method for backward failure propagation in conceptual system design

The results of our thematic analysis related to 'purpose' can be found in Table 6.2. The first column of the table provides high level purpose while the second column provides its sub-categories. The last column is an inventory of references whose title contains the theme in question.

Table 6.2 Themes related to 'Purpose' appearing in the titles of relevant ISFA documents

Category for purpose	Subcategory for purpose	References
Functional modeling		Chen et al. (2016), Zhao et al. (2019), Hunter et al. (2016)
Failures or faults	Failure or fault analysis	Mutha et al. (2013), Diao et al. (2018, 2021, 2022), Mansoor et al. (2023), Jensen et al. (2012), Keshavarzi et al. (2018), Sierla et al. (2013), Jiao et al. (2021), Papakonstantinou and Sierla (2012), Li et al. (2017a), Coatanéa et al. (2011), Mutha (2014); Ramos et al. (2020), Papakonstantinou et al. (2012a), Gunn (2022)
	Failures or faults modeling	Kapoor and Kumar (2014), O'Halloran (2013)
	Failures or faults reasoning	Jensen et al. (2014), Irshad et al. (2021)
	Failure/fault diagnosis	Li et al. (2022), Yang et al. (2018), Barbini et al. (2021), Li et al. (2016), Papakonstantinou et al. (2015)
	Failure/fault detection	Li et al. (2022), Papakonstantinou et al. (2014, 2015), Said et al. (2021), Jeya and Pillai (2012), Jeya and Singh (2013), Yahyaoui et al. (2021), Niculita et al. (2012)
	Failure/fault isolation	Niculita et al. (2012), Gonzalez et al. (2020)
	Failure classification	Thieme et al. (2020a)
	Failure interactions and emergence	Papakonstantinou et al. (2011, 2012b), Fan et al. (2016), Irshad (2021)
	Failure/fault assessment	Irshad et al. (2019)
Reliability		O'Halloran (2013), Sheetal and Taneja (2018), Sinha et al. (2019a, b, c, 2021), Lindén et al. (2016a, b), Zhu et al. (2021), Caroline and Sofianti (2018), Lin et al. (2014), Park et al. (2012), Zheng et al. (2023), Bharathi and Selvarani (2020), Smidts (2019), Ali et al. (2020)
Risk		Wang et al. (2021), Irshad (2021), Mimica et al. (2022), Nikula et al. (2015), Parhizkar et al. (2022), Thieme et al. (2020b), Papakonstantinou et al. (2013), Hu et al. (2022), Chen et al. (2018)

(continued)

Table 6.2 (continued)

Category for purpose	Subcategory for purpose	References
Safety		Mutha et al. (2011), Krishnan and Bhada (2020), Sierla et al. (2014), McIntire et al. (2016a), Krishnan and Bhada (2022), Tommila and Papakonstantinou (2016), Sierla et al. (2012)
Security		Jeya and Pillai (2012), Jeya and Singh (2013), Sierla et al. (2012)
Resilience		Mehrpouyan (2014)
Design		Huang et al. (2014), Keshavarzi (2018), Keshavarzi et al. (2017), McIntire et al. (2016b), Piacenza et al. (2017), Yu et al. (2016)
Cost		Kapoor and Kumar (2018a), Kumar and Kapoor (2013a, b)
Uncertainty Quantification		Hoyle et al. (2011)
Trade-off analysis		Barker et al. (2022a, b), McIntire (2016), Barker (2022)
Optimization		Li et al. (2022), Hoyle et al. (2014), Li et al. (2017b)
Analysis (Other)		Liu et al. (2022), Kapoor and Kumar (2018b), Wen-bo et al. (2012)
Decision-making		Jensen (2012), Piacenza et al. (2020)
Artificial intelligence		Li et al. (2017b), Mehrpouyan et al. (2014), Speith et al. (2023), Hayes et al. (2011), Liu et al. (2023), Makinson (2013)
Development		Bellman (2011), Aleem et al. (2018)
Management		Yang et al. (2021)
Testbeds		Pietrykowski (2022)

The same approach was taken regarding the theme of 'domain' or more specifically the object under study. This led to the results provided in Table 6.3. In the same fashion as in the table above, the first column is dedicated to high level categories of domains or objects of study while the second column explores sub-categories. The third column lists the references whose titles contain the category or sub-category.

In both tables we also highlighted with gray background areas ('purpose' categories or sub-categories' and 'domain' categories or sub-categories) where ISFA has already been applied. Other areas then are such that they could be explored for its further development as they have been deemed directly relevant in the literature.

Table 6.3 Themes related to 'Domain' appearing in the titles of relevant ISFA documents

Category for domain	Subcategory for domain	References
Systems	Large-scale systems	McIntire (2016)
	Complex systems	O'Halloran (2013), Barbini et al. (2021), Papakonstantinou et al. (2011, 2012b, 2013, 2014, 2015), Hu et al. (2022), McIntire et al. (2016a), Mehrpouyan (2014), Keshavarzi (2018), McIntire et al. (2016b), Piacenza et al. (2017, 2020), Yu et al. (2016), Hoyle et al. (2011), Mehrpouyan et al. (2014)
	Cyber-physical systems	Sierla et al. (2013), Papakonstantinou et al. (2012a), Ali et al. (2020), Hoyle et al. (2014)
	Multi-state systems	Chen et al. (2016), Zhao et al. (2019)
	Conceptual systems	Mansoor et al. (2023)
	Safety–critical systems	Diao et al. (2022), Papakonstantinou and Sierla (2012)
	With varying operating conditions	Sheetal and Taneja (2018), Lindén et al. (2016a)
	Resilient systems	Keshavarzi et al. (2017)
Product		Fan et al. (2016), Caroline and Sofianti (2018), Chen et al. (2018), Sierla et al. (2014), Hayes et al. (2011)
Foundational constituents of systems and their combinations	Hardware	Speith et al. (2023)
	Software	Mutha (2014), Jeya and Pillai (2012), Thieme et al. (2020a), Zhu et al. (2021), Park et al. (2012), Bharathi and Selvarani (2020), Thieme et al. (2020b), Mutha et al. (2011), Wang et al. (2015), Liu et al. (2022), Aleem et al. (2018)
	Human	Smidts (2019)
	Hardware and software	Yang et al. (2018), Sinha et al. (2019a, b, 2021), Zheng et al. (2023)
	Human, hardware, and software	Parhizkar et al. (2022)
	Human and system	Ramos et al. (2020)
	Network and architecture	Li et al. (2022), Tommila and Papakonstantinou (2016), Piacenza (2014)
Application domain	Energy	Diao et al. (2018), Wang et al. (2021), Li et al. (2016, 2017a), Mimica et al. (2022), Pietrykowski (2022)
	Space	Keshavarzi et al. (2018), Niculita et al. (2012), Huang et al. (2014), Wen-bo et al. (2012), Bellman (2011)

(continued)

Table 6.3 (continued)

Category for domain	Subcategory for domain	References
	Communications	Kapoor and Kumar (2014, 2018a, b); Gonzalez et al. (2020), Kumar and Kapoor (2013a, b)
	Internet of Things	Said et al. (2021), Yahyaoui et al. (2021)
	Robotics	Sierla et al. (2012)

Conclusion

This paper examines the various steps that were taken to develop the Integrated System Failure Analysis and the applications that were considered during its development. In addition, we identified areas that may offer potential for its future development and its applications.

References

Aleem S, Batool R, Ahmed F, Khattak AM (2018) Design guidelines for SaaS development process. In: 2018 IEEE 9th annual information technology, electronics and mobile communication conference (IEMCON). IEEE, pp 825–831

Ali N, Hussain M, Kim Y, Hong J-E (2020) A generic framework for capturing reliability in cyber-physical systems. In: Proceedings of the 2020 European symposium on software engineering, pp 148–153

Alidousti H, Taylor M, Bressloff NW (2011) Do capsular pressure and implant motion interact to cause high pressure in the periprosthetic bone in total hip replacement? J Biomechem Eng 133(12):121001-1-10

Barbini L, Bratosin C, Nägele T (2021) Embedding diagnosability of complex industrial systems into the design process using a model-based methodology. In: PHM society European conference, p 9

Barker TJ (2022) The impact of reliability in conceptual design-an integrated trade-off analysis. PhD Thesis, University of Arkansas

Barker TJ, Parnell GS, Pohl EA (2022) Integrating reliability in conceptual design trade-off analysis: a look at the literature. In: INCOSE international symposium, Wiley Online Library, pp 224–231

Barker T, Parnell GS, Pohl E, Specking E, Goerger SR, Buchanan RK (2022) Impact of reliability in conceptual design—An illustrative trade-off analysis. Systems 10(6):227

Bellman K (2011) Model-based design, engineering, and development: advancements mean new opportunities for space system development. In: AIAA SPACE 2011 conference & exposition, p 7304

Bharathi R, Selvarani R (2020) Hidden Markov model approach for software reliability estimation with logic error. Int J Autom Comput 17(2):305–320

Caroline H, Sofianti TD (2018) Enhancing efficiency of reliability assurance in product development through harmonization of methods: a case study at Dräger safety AG & Co. KGaA. PhD Thesis, Swiss German University

de Carvalho César Sobrinho ÁÁ and others (2016) Um método para o desenvolvimento e certificação de software de sistemas embarcados baseado em redes de petri coloridas e casos de garantia. Universidade Federal de Campina Grande

Chen Y, Zhao M, Huang J (2016) A state-behavior-function based approach for functional modeling of multi-state systems and its application. In: International design engineering technical conferences and computers and information in engineering conference American Society of Mechanical Engineers, p V007T06A030

Chen Z, He Y, Liu F, Zhu C, Zhou D (2018) Product infant failure risk modeling based on quality variation propagation and functional failure dependency. Adv Mech Eng 10(12):1687814018816587

Chow E, Willsky A (1984) Analytical redundancy and the design of robust failure detection systems. IEEE Trans Autom Contr 29(7):603–614

Coatanéa E, Nonsiri S, Ritola T, Tumer IY, Jensen DC (2011) A framework for building dimensionless behavioral models to aid in function-based failure propagation analysis

Diao X, Zhao Y, Pietrykowski M, Wang Z, Bragg-Sitton S, Smidts C (2018) Fault propagation and effects analysis for designing an online monitoring system for the secondary loop of the nuclear power plant portion of a hybrid energy system. Nucl Technol 202(2–3):106–123

Diao X, Smidts C, Mutha C (2021) Integrated system failure analysis software toolchain (IS-FAST). US 11138063 B1, October 05

Diao X, Pietrykowski M, Huang F, Mutha C, Smidts C (2022) An ontology-based fault generation and fault propagation analysis approach for safety-critical computer systems at the design stage. AI EDAM 36

Fan H, Liu Y, Cao Y, Qian B (2016) Efficient recognition of undesired coupling effects in system design of multidisciplinary products. J Eng Des 27(10):665–696

Forbus KD (1988) Qualitative physics: past, present, and future. In: Exploring artificial intelligence. Elsevier, pp 239–296

Gonzalez AJ et al (2020) The isolation concept in the 5G network slicing. In: 2020 European conference on networks and communications (EuCNC). IEEE, pp 12–16

Gunn CA (2022) Quantifying consequences of externally induced failures propagated through systems during functional system design. PhD Thesis, Monterey, CA, Naval Postgraduate School

Harzing AW (2007) Publish or Perish. https://harzing.com/resources/publish-or-perish

Hayes CC, Goel AK, Tumer IY, Agogino AM, Regli WC (2011) Intelligent support for product design: looking backward, looking forward

Hoyle C, Tumer IY, Kurtoglu T, Chen W (2011) Multi-stage uncertainty quantification for verifying the correctness of complex system designs. In: International design engineering technical conferences and computers and information in engineering conference, pp 1169–1178

Hoyle C, Piacenza J, DuPont B, Cotilla-Sanchez E (2014) Robust optimization of complex cyber-physical systems. In: Proceedings of the international annual conference of the american society for engineering management, American Society for Engineering Management (ASEM), p 1

Hu Y, Parhizkar T, Mosleh A (2022) Guided simulation for dynamic probabilistic risk assessment of complex systems: concept, method, and application. Reliab Eng Syst Saf 217:108047. https://doi.org/10.1016/j.ress.2021.108047

Huang W, Zhang W, Chen L, Shi S, Cai Y (2014) Research on spacecraft design for ORS based on the systems theory. Proc Inst Mech Eng, Part G: J Aerosp Eng 228(6):949–959

Hunter SC, Jensen DC, Tumer IY, Hoyle C (2016) The impact of abstraction and fidelity levels on the usefulness of early system functional models. In: International design engineering technical conferences and computers and information in engineering conference. American Society of Mechanical Engineers, p V01BT02A018

Irshad L (2021) A framework to evaluate the risk of human-and component-related vulnerability interactions

Irshad L, Ahmed S, Demirel O, Tumer IY (2019) Coupling digital human modeling with early design stage human error analysis to assess ergonomic vulnerabilities. In: AIAA SciTech 2019 forum, p 2349

Irshad L, Demirel HO, Tumer IY (2021) The human error and functional failure reasoning framework: how does it scale? In: International design engineering technical conferences and computers and information in engineering conference. American Society of Mechanical Engineers, p V002T02A021

Jensen DC (2012) Enabling safety-informed design decision making through simulation, reasoning and analysis. Oregon State University

Jensen DC, Hoyle C, Tumer IY (2012) Clustering function-based failure analysis results to evaluate and reduce system-level risks. In: International design engineering technical conferences and computers and information in engineering conference. American Society of Mechanical Engineers, pp 1055–1064

Jensen DC, Bello O, Hoyle C, Tumer IY (2014) Reasoning about system-level failure behavior from large sets of function-based simulations. AI EDAM 28(4):385–398

Jeya S, Pillai SMP (2012) Intrusion detection system for relational databases. i-Manager's J Softw Eng 6(4):9

Jeya S, Singh TJJ (2013) Intrusion detection system using binary classifier Algorithm. i-Manager's J Softw Eng 7(3):21

Jiao J, Pang S, Chu J, Jing Y, Zhao T (2021) An improved FFIP method based on mathematical logic and SysML. Appl Sci 11(8):3534

Kapoor S, Kumar R (2014) Comparative analysis of two stochastic models for a base transceiver system considering hardware and software interaction failures. Arya Bhatta J Math Inf 6(2):313–322

Kapoor S, Kumar R (2018a) Comparative cost-benefit analysis of two reliability models for one unit base transceiver system considering hardware based software faults. Int J Stat Appl Math 1(3):278–286

Kapoor S, Kumar R (2018b) Stochastic analysis of a base transceiver system considering traffic congestion and chances of hardware/software expansions. Int J Oper Res 32(3):364–379

Keshavarzi E (2018) Resilient design for complex engineered systems in the early design phase

Keshavarzi E, McIntire M, Goebel K, Tumer IY, Hoyle C (2017) Resilient system design using cost-risk analysis with functional models. In: International design engineering technical conferences and computers and information in engineering conference. American Society of Mechanical Engineers, p V02AT03A043

Keshavarzi E, Goebel K, Tumer I, Hoyle C (2018) Failure analysis in conceptual phase toward a robust design: case study in monopropellant propulsion system. Int J Res Eng 5(9):535–546

Kumar R, Kapoor S (2013a) Economic and performance evaluation of stochastic model on a base transceiver system considering various operational modes and catastrophic failures. J Math Stat 9(3):198–207

Kumar R, Kapoor S (2013b) Profit evaluation of a stochastic model on base transceiver system considering software based hardware failures and congestion of calls. Int J Appl Innov Eng Manag 2(3):554–562

Krishnan R, Bhada SV (2020) An integrated system design and safety framework for model-based safety analysis. IEEE Access 8:146483–146497

Krishnan R, Bhada SV (2022) Integrated system design and safety framework for model-based safety assessment. IEEE Access 10:79311–79334
Li H, Diao X, Li B, Smidts C, Bragg-Sitton S (2017a) fault propagation and effects analysis for designing an online monitoring system for the secondary loop of a nuclear power plant part of a hybrid energy system. Idaho National Lab.(INL), Idaho Falls, ID (United States)
Li Y, Sun B, Wang Z, Ren Y (2017b) Ontology-based environmental effectiveness knowledge application system for optimal reliability design. J Comput Inf Sci Eng 17(1):011005
Li B, Diao X, Vaddi PK, Gao W, Smidts C (2022) A propagation-based fault detection and discrimination method and the optimization of sensor deployment. Ann Nucl Energy 166:108746
Li H, Bragg-Sitton S, Smidts C (2016) Failure diagnosis for the holdup tank system via ISFA. Idaho National Lab.(INL), Idaho Falls, ID (United States)
Lin PT, Chou Y-C, Ting Y, Shyu S-S, Chen C-K (2014) A robust system reliability analysis using partitioning and parallel processing of Markov chain. AI EDAM 28(4):311–322
Lindén J, Söderberg A, Sellgren U (2016a) Reliability assessment with varying operating conditions. Procedia CIRP 50:796–801
Lindén J, Sellgren U, Söderberg A (2016b) Model-based reliability analysis. AI EDAM 30(3):277–288
Liu H, Jin Z, Zheng Z, Huang C, Zhang X (2022) An ontological analysis of safety-critical software and its anomalies. In: 2022 IEEE 22nd international conference on software quality, reliability and security (QRS), IEEE, pp 311–320
Liu Z, Zhang X, Khanduri P, Lu S, Liu J (2023) Prometheus: taming sample and communication complexities in constrained decentralized stochastic bilevel learning
Makinson KA (2013) Preliminary framework for the run-ahead predictive simulation software (RAPSS). Oregon State University
Mansoor A, Diao X, Smidts C (2023) A method for backward failure propagation in conceptual system design. Nuclear Sci Eng 1–27. https://doi.org/10.1080/00295639.2023.2196937
McIntire MG (2016) From functional modeling to optimization: risk and safety in the design process for large-scale systems
McIntire MG, Keshavarzi E, Tumer IY, Hoyle C (2016) Functional models with inherent behavior: towards a framework for safety analysis early in the design of complex systems. In: ASME international mechanical engineering congress and exposition. American Society of Mechanical Engineers, p V011T15A035
McIntire MG, Hoyle C, Tumer IY, Jensen DC (2016b) Safety-informed design: using subgraph analysis to elicit hazardous emergent failure behavior in complex systems. AI EDAM 30(4):466–473
Mehrpouyan H (2014) A framework for assessing and improving the resilience of complex engineered systems during the early design process
Mehrpouyan H, Tumer IY, Hoyle C, Giannakopoulou D, Brat G (2014) Formal verification of complex systems based on sysml functional requirements. In: Annual conference of the PHM society
Mimica M, De Urtasun LG, Krajačić G (2022) A robust risk assessment method for energy planning scenarios on smart islands under the demand uncertainty. Energy 240:122769
Mutha CV (2014) Software fault propagation and failure analysis for UML based software design. PhD Thesis, The Ohio State University
Mutha C, Smidts C (2011) An early design stage UML-based safety analysis approach for high assurrance software systems. In: 2011 IEEE 13th international symposium on high-assurance systems engineering. IEEE, pp 202–211
Mutha C, Jensen D, Tumer I, Smidts C (2013) An integrated multidomain functional failure and propagation analysis approach for safe system design. AIEDAM 27(4):317–347. https://doi.org/10.1017/S0890060413000152
Niculita I-O, Irving P, Jennions IK (2012) Use of COTS functional analysis software as an IVHM design tool for detection and isolation of UAV fuel system faults

Nikula H, Sierla S, O'Halloran B, Karhela T (2015) Capturing deviations from design intent in building simulation models for risk assessment. J Comput Inf Sci Eng 15(4):041011
O'Halloran BM (2013) A framework to model reliability and failures in complex systems during the early engineering design process. Oregon State University
Papakonstantinou N, Sierla S, Jensen DC, Tumer IY (2011) Capturing interactions and emergent failure behavior in complex engineered systems at multiple scales. In: International design engineering technical conferences and computers and information in engineering conference, pp 1045–1054
Papakonstantinou N, Sierla S (2012) Early phase fault propagation analysis of safety critical factory automation systems. In: IEEE 10th international conference on industrial informatics. IEEE, pp 364–369
Papakonstantinou N, Sierla S, Tumer IY, Jensen DC (2012a) Using fault propagation analyses for early elimination of unreliable design alternatives of complex cyber-physical systems. In: International design engineering technical conferences and computers and information in engineering conference. American Society of Mechanical Engineers, pp 1183–1191
Papakonstantinou N, Sierla S, Jensen DC, Tumer IY (2012b) Simulation of interactions and emergent failure behavior during complex system design. J Comput Inf Sci Eng 12(3):031007
Papakonstantinou N, Sierla S, O'Halloran B, Tumer IY (2013) A simulation based approach to automate event tree generation for early complex system designs. In: International design engineering technical conferences and computers and information in engineering conference. American Society of Mechanical Engineers, p V02BT02A008
Papakonstantinou N, Proper S, O'Halloran B, Tumer IY (2014) Simulation based machine learning for fault detection in complex systems using the functional failure identification and propagation framework. In: International design engineering technical conferences and computers and information in engineering conference. American Society of Mechanical Engineers, p V01BT02A022
Papakonstantinou N, Proper S, O'Halloran B, Tumer IY (2015) A plant-wide and function-specific hierarchical functional fault detection and identification (HFFDI) system for multiple fault scenarios on complex systems. In: International design engineering technical conferences and computers and information in engineering conference. American Society of Mechanical Engineers, p V01BT02A039
Park J, Kim H-J, Shin J-H, Baik J (2012) An embedded software reliability model with consideration of hardware related software failures. In: 2012 IEEE sixth international conference on software security and reliability. IEEE, pp 207–214
Parhizkar T, Utne IB, Vinnem J-E, Parhizkar T, Utne IB, Vinnem J-E (2022) Human, hardware, and software interactions in risk assessment. In: Online probabilistic risk assessment of complex marine systems: principles, modelling and applications, pp 55–74
Piacenza III JR (2014) Design of robust infrastructure systems incorporating user behavior
Piacenza JR, Proper S, Bozorgirad MA, Hoyle C, Tumer IY (2017) Robust topology design of complex infrastructure systems. ASCE-ASME J Risk Uncertainty Eng Syst, Part B: Mech Eng 3(2):021006
Piacenza JR, Faller KJ, Bozorgirad MA, Cotilla-Sanchez E, Hoyle C, Tumer IY (2020) Understanding the impact of decision making on robustness during complex system design: More resilient power systems. ASCE-ASME J Risk Uncertainty Eng Syst, Part B: Mech Eng 6(2):021001
Pietrykowski MC (2022) Experimental test facility framework for nuclear applications, PhD Thesis. The Ohio State University
Ramos MA, Thieme CA, Utne IB, Mosleh A (2020) A generic approach to analysing failures in human–system interaction in autonomy. Saf Sci 129:104808

Said AM, Yahyaoui A, Abdellatif T (2021) Efficient anomaly detection for smart hospital IoT systems. Sensors 21(4):1026

Saldaña J (2011) The coding manual for qualitative researchers. Sage

Sheetal DS, Taneja G (2018) Reliability analysis of a system working in high temperature zones with fault-dependent repair during night hours. Int J Appl Eng Res 13(20):14650–14656

Shine R et al (2017) Modeling of biodegradable polyesters with applications to coronary stents. J Med Devices 11(2):021007

Sierla S, Tumer I, Papakonstantinou N, Koskinen K, Jensen D (2012) Early integration of safety to the mechatronic system design process by the functional failure identification and propagation framework. Mechatronics 22(2):137–151

Sierla S, O'Halloran BM, Karhela T, Papakonstantinou N, Tumer IY (2013) Common cause failure analysis of cyber–physical systems situated in constructed environments. Res Eng Design 24:375–394

Sierla S, O'Halloran BM, Nikula H, Papakonstantinou N, Tumer IY (2014) Safety analysis of mechatronic product lines. Mechatronics 24(3):231–240

Sinha S, Goyal NK, Mall R (2019a) Early prediction of reliability and availability of combined hardware-software systems based on functional failures. J Syst Architect 92:23–38

Sinha S, Goyal NK, Mall R (2019b) Early prediction of reliability/availability for embedded system based on conceptual design

Sinha S, Goyal NK, Mall R (2019c) Survey of combined hardware–software reliability prediction approaches from architectural and system failure viewpoint. Int J Syst Assur Eng Manag 10:453–474

Sinha S, Goyal NK, Mall R (2021) Reliability and availability prediction of embedded systems based on environment modeling and simulation. Simul Model Pract Theory 108:102246

Smidts C (2019) Human reliability as a science—A divergence on models. In: Risk based technologies, pp 127–142

Speith T, Speith J, Becker S, Zou Y, Biega A, Paar C (2023) Expanding explainability: from explainable artificial intelligence to explainable hardware. arXiv preprint. arXiv:2302.14661

Thieme CA, Mosleh A, Utne IB, Hegde J (2020a) Incorporating software failure in risk analysis–Part 1: software functional failure mode classification. Reliab Eng Syst Saf 197:106803

Thieme CA, Mosleh A, Utne IB, Hegde J (2020b) Incorporating software failure in risk analysis—-Part 2: risk modeling process and case study. Reliab Eng Syst Saf 198:106804

Tommila T, Papakonstantinou N (2016) Challenges in defence in depth and I&C architectures. VTT Research Report

Wang Q, Diao X, Zhao Y, Chen F, Yang G, Smidts C (2021) An expert-based method for the risk analysis of functional failures in the fracturing system of unconventional natural gas. Energy 220:119570. https://doi.org/10.1016/j.energy.2020.119570

Wang X, Zhang K, Wu Q (2015) A design of security assessment system for e-commerce website. In: 2015 8th international symposium on computational intelligence and design (ISCID). IEEE, pp 137–140

Wen-bo H, Wei-hua Z, Ye-quan C, Shuai S (2012) Systems analysis on spacecraft design. In: 2012 3rd international conference on system science, engineering design and manufacturing informatization, IEEE, pp 197–200

Yahyaoui A, Abdellatif T, Yangui S, Attia R (2021) READ-IoT: reliable event and anomaly detection framework for the Internet of Things. IEEE Access 9:24168–24186

Yang J, Aldemir T, Smidts C (2018) A deductive method for diagnostic analysis of digital instrumentation and control systems. IEEE Trans Reliab 67(4):1442–1458

Yang C, Quan L, Liao L (2021) Intelligent decision techniques for construction engineering management research: a science mapping analysis and future trends. In: International symposium on advancement of construction management and real estate. Springer, pp 721–736

Yu BY, Honda T, Zubair SM, Sharqawy MH, Yang MC (2016) A maintenance-focused approach to complex system design. AI EDAM 30(3):263–276

Zhao M, Chen Y, Chen L, Xie Y (2019) A state–behavior–function model for functional modeling of multi-state systems. Proc Inst Mech Eng C J Mech Eng Sci 233(7):2302–2317

Zheng Z, Yang J, Huang J (2023) Software-hardware embedded system reliability modeling with failure dependency and masked data. SSRN 4502314

Zhu J, Gong Z, Sun Y, Dou Z (2021) Chaotic neural network model for SMISs reliability prediction based on interdependent network SMISs reliability prediction by chaotic neural network. Qual Reliab Eng Int 37(2):717–742

박진희, 백종문, and 신주환 (2011) 하드웨어와소프트웨어의상호작용을고려한시스템신뢰성모델링접근방법. 한국정보과학회학술발표논문집 38(2B):147–150

王恩亮 and 张丽华 (2014) "嵌入式系统软件可靠性模型的研究. 佳木斯大学学报 (自然科学版) 32(6):873–875

陆寅, 秦树东, 郭鹏, and 董云卫, "软硬件综合 AADL 可靠性建模及分析方法 (形式化方法与应用)," 软件学报

Chapter 7
Digital Twins: Definition, Implementation and Applications

Diego Galar and Uday Kumar

Introduction

The digital technologies accompanying Industry 4.0 have ushered in a new era in the management of industrial economic systems. The concept of the digital twin is at the heart of this transformation. Stemming from the convergence of advanced data analytics, Internet of Things (IoT) technologies, and virtual modelling and domain knowledge (Fig. 7.1), digital twins were conceptualized to create virtual replicas of physical assets and systems.

Digital twin technology allows real-time monitoring, analysis, and simulation of industrial operations, leading to enhanced predictive maintenance, optimized production workflows, and improved product development. It facilitates a comprehensive understanding of complex industrial systems, enabling precise insights into performance, functionality, and potential areas for optimization. Given their ability to simulate and anticipate various scenarios, digital twins have become instrumental in driving innovation and efficiency, providing a solid foundation for the transformative journey toward an interconnected and intelligent industrial landscape. This innovative technology offers a comprehensive understanding of the unique attributes, operational performance, and potential issues of any equipment or system. Notably, the digital twin facilitates the virtual training of operators, eliminating the need for dedicated trainers or simulators.

With the continuous advancement of machine learning (ML) and Artificial Intelligence (AI), the realm of autonomous industrial machines is poised to undergo a significant shift. In this autonomous landscape, the role of the digital twin will evolve, propelling machines toward increased self-awareness and autonomy. Equipped with

D. Galar (✉) · U. Kumar
Lulea University of Technology, Lulea, Sweden
e-mail: diego.galar@ltu.se

U. Kumar
e-mail: uday.kumar@ltu.se

P. V. Varde et al. (eds.), *Advances in Risk-Informed Technologies*, Risk, Reliability and Safety Engineering, https://doi.org/10.1007/978-981-99-9122-8_7

Fig. 7.1 Digital twins as cocreation of O&M knowledge and AI

the capability to optimize their own performance, synchronize with other machines, conduct self-diagnosis, and autonomously rectify faults, machines will necessitate minimal intervention from human operators (Happiest Minds 2021).

The digital twin represents the convergence of the physical and virtual worlds, combining various technologies such as AI, ML, and software analytics to create dynamic digital simulation models. These models continuously update and adapt to reflect changes in their physical counterparts. By providing a precise digital replica of machinery, the digital twin technology enables operators to gain insights into the distinctive characteristics of the machine, its operational efficiency, and potential issues. Real-time monitoring through sensors enables operators to receive timely alerts about potential failures, downtimes, or accidents, allowing them to optimize the machine's performance, monitor inter-device coordination, diagnose issues, and rectify faults with minimal impact on productivity.

This evolving landscape of system development and management is witnessing a significant shift towards making systems and system-of-systems smarter using digital twin technologies. This transformation is driven by the integration of cutting-edge technologies, including IoT and user-friendly interfaces, which have revolutionized system interaction and decision-making processes.

A fundamental aspect of this transformation is Model-Based Systems Engineering (MBSE), a concept that emphasizes system reuse throughout its lifecycle. MBSE facilitates communication among stakeholders and is incorporated early in the acquisition process to streamline system synthesis. A system specification plays a crucial role in defining the requirements of a technical or software system under development. MBSE takes this concept further by formalizing and consistently applying modeling techniques throughout the system's lifecycle, from its conceptual phase to design and beyond. MBSE supports various aspects of system development, including requirements, architecture, analysis, verification, and validation. By employing formal and model-based specification techniques, it simplifies the process of specifying complex systems. The key to MBSE is the creation and utilization of a coherent digital system model, which serves as the central source of all pertinent information, streamlining interdisciplinary specification and development processes.

Formal and semi-formal modeling languages are employed to concisely represent the system's requirements, structure, and behavior.

While some publications explore the use of digital twins within a model-based product development framework or the integration of MBSE into digital twins through diverse methods, the literature does not extensively focus on employing MBSE to manage digital twin complexity or to specify digital twins themselves. Some publications highlight specific advantages of MBSE for digital twin development but may not provide a comprehensive specification technique. They often emphasize the requirements of particular stakeholders or overlook the entire product lifecycle. However, some work is starting to address the need for a holistic framework for digital twins, recognizing research gaps related to considering the full lifecycle and identifying the requirements of various stakeholders throughout the lifecycle (Fig. 7.2) (Rasor 2021).

This concept aligns with the broader digital transformation initiatives pursued by many companies and government agencies. It empowers stakeholders across development, operations, and support with accessible and standardized data, thereby enhancing decision-making processes. In a comprehensive system engineering approach to digital twins, the digital twin itself is treated as a distinct system, resulting in a system-of-systems framework. This approach introduces a new paradigm of integrated system design and modelling.

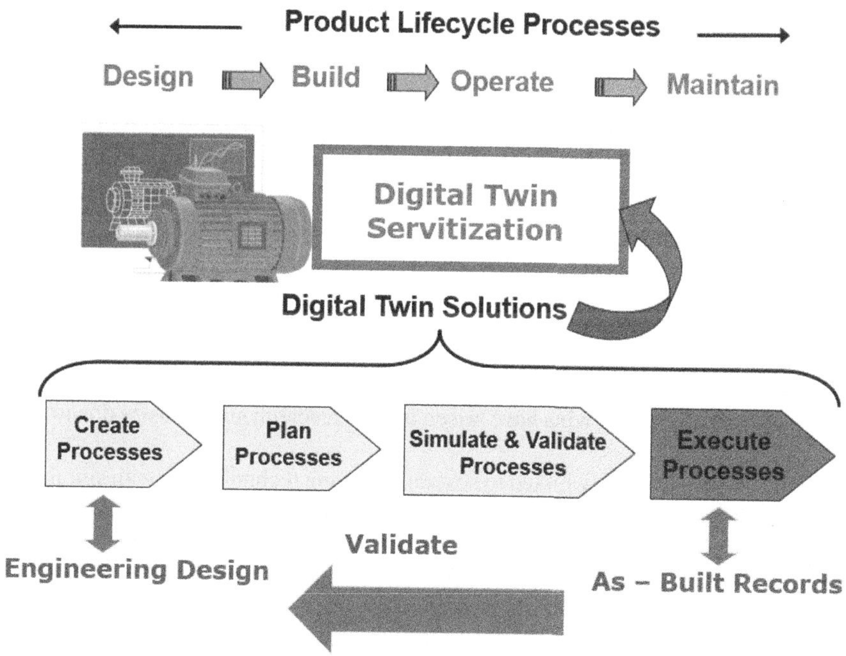

Fig. 7.2 Digital twins and the whole lifecycle dimension

The increasing relevance of Industry 4.0 and the Industrial Internet of Things (IIoT) has expanded the scope of digital twin applications. Notably, digital twin research focuses on creating purpose-oriented virtual models that represent physical systems. Other efforts are directed at establishing bidirectional relationships between physical objects and their virtual counterparts, facilitating data transfer and processing.

Various terminologies are used to describe digital concepts, such as the Asset Administration Shell in the context of Platform Industry 4.0. Related terms include Digital Model, Digital Master, and Digital Shadow, each serving distinct purposes in connecting physical and virtual realms.

The multitude of digital twin research approaches has led to diverse understandings and use cases, often lacking explicit specifications of required resources. Moreover, there is a growing need to determine the added value of implementing specific use cases. At the moment, there is no holistic framework for the conception, development, and implementation of digital twins; thus, there is a need or further exploration and standardization (Rasor 2021).

Digital Twin Definition

The concept of the digital twin has gained significant traction in the era of Industry 4.0, but there are a number of different definitions and interpretations. Some view digital twins as digital representations of physical objects or systems from the real world, while others consider them realistic digital depictions of physical entities. In essence, a digital twin encompasses a comprehensive description of a component, product, or system, containing all relevant information for its current and future lifecycle phases.

A digital twin is essentially a virtual model intricately integrated with its real-world counterpart. However, digital twins can vary significantly in terms of detail, technical focus, and scope. They have emerged as a critical technology in modern design and production engineering workflows, driven by advancements in sensor technology, information systems, and simulation technologies like Cyber-Physical Systems (CPS) and the Industrial Internet of Things (IIoT). Different interpretations and definitions of digital twins have arisen in both research and industry due to their diverse application areas.

Digital twins are closely linked to several emerging technologies, with simulation systems, communication technologies, and CPS playing pivotal roles. CPS, in particular, represent a fundamental concept in Industry 4.0 and are a technical evolution of mechatronic systems that blend mechanics, electronics, and computer science. CPS are equipped with sensors for data collection, actuators for interacting with their surroundings, and embedded systems, which are microcomputers with computing capabilities and unique identities. They can communicate and coordinate with each other via data infrastructure, typically the Internet, creating cyber-physical production systems (CPPS) when deployed in a production environment (Bauer 2015).

Creating a digital twin relies on decentralized data collection and processing by CPS, using data from multiple CPS. This process involves addressing challenges related to data acquisition, transfer, storage, security, and analysis, and it requires a combination of dedicated hardware and software solutions. The adoption of the 5G communication standard is anticipated to address current limitations in terms of bandwidth, latency, resilience, and scalability, particularly when supporting multiple devices.

Once data from CPS are gathered, digital twins facilitate the running of simulations to explore various scenarios, aiding in predicting the behavior of CPS. Some experts have even asserted that from a simulation perspective, the digital twin approach represents the next significant advancement in modeling, simulation, and optimization technology.

In summary, CPS comprise a conceptual framework and technology for smartness, and digital twins underpin the infrastructure of Industry 4.0. CPS serves as a fundamental framework that combines computing elements and physical processes, enabling the seamless integration of the digital and physical worlds. This integration forms the backbone for the development of intelligent and interconnected technologies, known as smart systems, which encompass automation, data-driven decision-making, and adaptive functionalities. Digital twins, in turn, leverage the capabilities of CPS to create virtual replicas that mimic the behavior of physical assets or processes in real-time. By utilizing data collected from a network of sensors and IoT devices, digital twins facilitate real-time monitoring, analysis, and optimization of complex industrial systems. The interwoven relationship between CPS, smart technologies, and digital twins fosters enhanced operational efficiency, predictive maintenance, and overall system resilience, driving the transformation of modern industries towards an interconnected and intelligent future (Fig. 7.3).

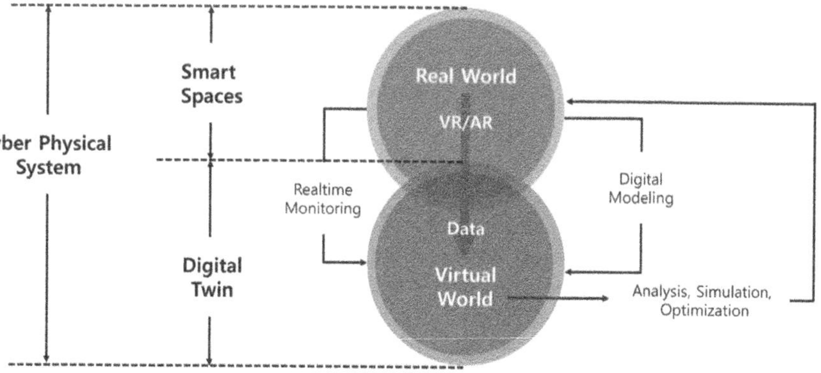

Fig. 7.3 Digital twin as natural outcome of CPSs

Simulation, in general, involves replicating the operation of real-world processes or systems, typically focusing on the evolution of physical quantities or entities of interest over time, across various physical domains. Simulation models describe mathematical, logical, and symbolic relationships among these entities, and these relationships can vary based on the intended use of the model.

Throughout the product lifecycle, different stages can be identified, and numerous simulation technologies have emerged over the years to address each of these stages. These simulation tools continue to evolve, offering increased fidelity and enabling a deeper understanding of how design decisions impact product behavior in real-world use. It's important to note that a digital twin isn't a single, all-encompassing model but rather a collection of interconnected operational data artifacts and simulation models. These models must be chosen with the appropriate level of granularity for their intended purposes and evolve throughout the product lifecycle. For example, simpler models may be suitable for conceptual product decisions, while more sophisticated simulations support detailed product design and manufacturing processes.

Digital twins generate vast amounts of data, necessitating robust data processing methods. AI models, which leverage ML techniques like neural networks, have become increasingly powerful thanks to enhanced computing capabilities. AI models can be deployed in cloud or distributed computing environments or embedded directly in physical objects like robots and vehicles to ensure data security and enable local processing of sensitive information. In distributed systems, ensuring data integrity is paramount, and blockchain technology can provide solutions for data protection and traceability of events throughout the product lifecycle. Blockchain can also facilitate the use of smart contracts, small software components that can automate actions such as maintenance or supply chain transactions within the digital twin ecosystem (Dittrich 2019).

Origin and History of Digital Twins

The concept of the digital twin has intriguing historical roots in NASA's Apollo project. During this project, a physical space capsule on Earth was used to simulate the behaviour of a similar capsule in space. While this example involved a physical representation, it captures the essence of having one object mimic the effects of another. However, the space capsule on Earth was not a digital representation.

Following NASA's lead, the US Air Force embraced digital technology for various purposes, including design, maintenance, and failure prediction. The goal was to use digital twins to simulate the physical and mechanical properties of aircraft to predict issues like fatigue or cracks, ultimately extending the remaining useful life (RUL) of these assets. As the concept of digital twins gained momentum, it found applications in sustainable space exploration and the design of aerospace vehicles, marking its continued evolution and relevance in various industries (Singh 2021).

Digital twins have various precursors leading to their modern incarnation:

- Mirror Worlds (1991): David Gelernter proposed the concept of "Mirror Worlds," where software models would replicate reality based on information from the physical world.
- Mirrored Spaces Model (2002): Michael Grieves introduced a model featuring real space, virtual space, and a linking mechanism to exchange data between them.
- Information Mirroring Model (2006): Grieves refined his model and renamed it the "Information Mirroring Model." This model introduced bidirectional linking between real and virtual spaces and allowed for multiple virtual spaces corresponding to a single real space, enabling the exploration of alternate ideas or designs.
- Digital Shadow and Digital Model: A digital model represents a physical object but involves only manual data exchange. It lacks real-time synchronization with the physical object. A digital shadow is a static copy of the physical object's data, with one-way data flow from the physical object to its digital representation. It does not reflect the real-time state of the physical object.
- Semantic Virtual Factory Data Model: The model represents virtual entities within a factory environment, primarily used in manufacturing and industrial contexts. Unlike digital twin, it focuses on data modelling alone and does not offer real-time synchronization with physical objects.
- Product Avatar: Product avatar is a distributed and decentralized approach to managing product information. However, it lacks the concept of feedback and may provide information on only specific parts of a product.
- Digital Product Memory: Digital product memory involves sensing and capturing information related to specific physical parts or products. It was a precursor to the broader capabilities of digital twin.
- Intelligent Product: The concept of an intelligent product incorporates technologies like IoT, Big Data, and ML but lacks the comprehensive integration and synchronization offered by digital twin. Digital twin builds upon the foundation of intelligent products.
- Holons: These are early computer-integrated manufacturing tools that laid the groundwork for subsequent technologies. They contributed to the development of concepts like digital twin.

Despite these early conceptualizations, practical implementation of digital twins faced significant challenges due to limitations in technology. Factors such as low computing power, limited device connectivity to the internet, inadequate data storage and management, and underdeveloped machine algorithms hindered the practical application of digital twins during this period. The concept of the digital twin evolved significantly with the rise of IoT, a fundamental component of Industry 4.0.

Figure 7.4 illustrates the progression of digital twin concepts over time.

Today, there is no universally accepted standard definition for the term "digital twin". Instead, various definitions have emerged based on specific characteristics that stem from different use cases involving digital twins. A common thread among these definitions is the integration of diverse data sources to create a digital representation of

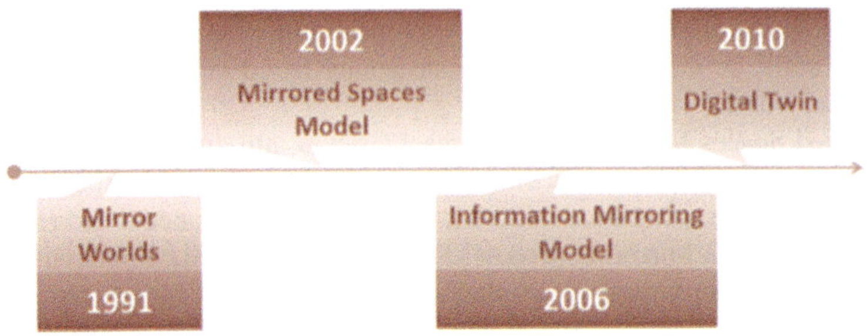

Fig. 7.4 Timeline of evolution of digital twin (Singh 2021)

a physical object or process throughout its entire lifecycle. This digital representation serves as the foundation for conducting various analyses and simulations (Van der Valk 2020).

A digital twin serves as a virtual model of a real-world system, process, or service and can be applied to model products, factories, or business services. It offers the capability for real-time monitoring of systems and processes, enabling timely data analysis to prevent issues before they arise, schedule preventative maintenance, minimize or prevent downtimes, explore new business opportunities, and plan future updates and innovations. While traditional virtual models often represent general concepts of a system or its components, a digital twin is an instance, a specific representation of a real-world counterpart. Digital twin technology can reduce the cost of system verification and testing while providing a real-time assessment of the system's performance.

In summary, the digital twin concept represents a significant advancement over its predecessors. It combines real-time synchronization, comprehensive data exchange, and feedback mechanisms between the physical and digital worlds. While earlier concepts served specific purposes, the digital twin integrates these functionalities to create a holistic and dynamic representation of physical objects, making it a valuable tool across various industries and applications.

Defining the Digital Twin

As mentioned above, the term "digital twin" is relatively new and has various interpretations and definitions depending on the context and organization. Different entities have their own perspectives on what a digital twin represents:

- General Electric (GE) refers to digital twins as "dynamic digital models of physical assets and systems."
- Siemens defines digital twins as "a digital copy that is created and developed simultaneously with the real machine."
- DNV GL describes digital twins as "a virtual image of an asset, maintained throughout the lifecycle and easily accessible at any time."
- SAP defines digital twins as digital representations that use real-time data from sensors to continuously represent a physical reality.

While these definitions vary, they all share common intrinsic characteristics that define what a digital twin is:

Identity: A digital twin is always associated with a real-world object or system. It represents a one-to-one or one-to-many mapping between the object or system and its digital twin counterpart.

Representation: A digital twin captures the essential physical manifestation of the real asset in a digital format, which can include computer aided design (CAD) or computer aided engineering (CAE) models with corresponding metadata.

(1) *State*: Unlike traditional CAD/CAE models, a digital twin has the capability to render quantifiable measures of the asset's state in close to real-time.
(2) *Behavior*: A digital twin reflects basic responses to external stimuli, such as forces, temperatures, or chemical processes, within its operational context.
(3) *Context*: A digital twin describes the external operating context in which the asset exists or operates, including factors like wind, waves, temperature, and more.

These characteristics ensure that digital twins offer a genuine view of the virtual system and its real-world status. When the model corresponds uniquely to an identifiable object and accurately reflects its state, it qualifies as a digital twin. Furthermore, in some cases, digital twins can even initiate operational changes in the physical object they represent (Makarov 2019).

Benefits of Digital Twins

Digital twins offer several advantages, including:

- Monitoring and inspection: Digital twins enable monitoring and inspection of assets digitally, saving effort and resources compared to physical inspections, especially in challenging access scenarios.
- Data aggregation: Digital twins facilitate high-fidelity data aggregation, such as stress cycle counting in fatigue life utilization calculations.
- Remaining life assessment: Digital twins can assess the remaining life of structures, aiding in maintenance and longevity.
- Early damage detection: Digital twins can detect damage early, enabling pre-emptive maintenance and preventing shutdowns.
- Design feedback: Digital twins provide access to aggregated time series data for design feedback, transitioning from hindsight to foresight.
- Visualization and stress analysis: Digital twins allow visualization and inspection of stresses at inaccessible or hidden locations.

As technology advances, digital twins are becoming increasingly sophisticated, with high-definition maps and detailed mathematical models of physical objects. These digital twins have the potential to revolutionize various industries, from self-driving cars with highly precise maps to regulatory reviews of medical devices based on realistic mathematical models.

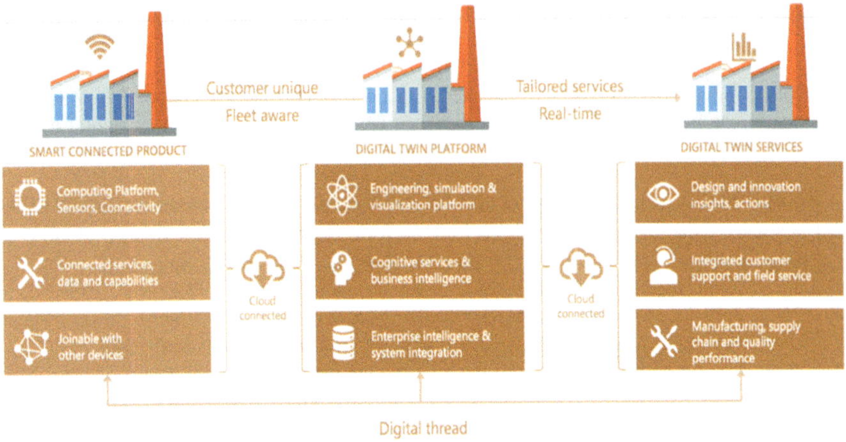

Fig. 7.5 Architecture of digital twins

Digital Twin Architecture

The digital twin architecture links physical and virtual worlds (Juarez 2021):

1. Physical World: This element encompasses the tangible, real-world entities targeted for replication or modeling by the digital twin. In manufacturing, these physical entities may encompass machines, equipment, assets, products, and the entire production environment. The physical world (see Fig. 7.5) includes devices and sensors:

- Devices: These are the physical objects or assets themselves, such as machines or equipment used in manufacturing processes.
- Sensors: Sensors represent physical components directly connected to devices, responsible for collecting real-time data and information from the physical world. Sensors capture essential data, which is subsequently transmitted to the digital world for processing.

2. Digital world: the digital world comprises two essential components:

- Virtual environment platform (VMP): The VMP serves as an integrated 3D digital model capable of executing applications and actions to validate various algorithms. It provides the foundation for creating and operating digital twins, offering the requisite models for their effective development and utilization. It can be considered middleware that links (see Fig. 7.5) smartness with delivered services.
- Digital twins: Digital twins are virtual representations of their corresponding physical objects. They faithfully mirror the life cycle and behavior of these physical entities, enabling a wide array of operations, including control, prediction, and analysis.

3. Connections between physical and digital worlds: These connections facilitate the exchange of data and information between the real and virtual domains. The nature of these connections may vary depending on the specific development methodology employed. They are vital for ensuring that the digital twin accurately reflects the state and behavior of the physical object.

These components collectively form the core of the digital twin concept. They enable organizations to create virtual counterparts of physical assets and systems, empowering real-time monitoring, analysis, and informed decision-making. This capability holds substantial potential for enhancing efficiency and effectiveness across various domains, including manufacturing.

Digital Twin Classes and Categories

Digital twins are applied across domains, playing pivotal roles in decision-making, real-time monitoring, and behavior prediction for tangible objects. The primary digital twin classes include:

- Digital twin of products: Originally developed for aerospace applications, this class manages data related to specific product lifecycles. Sensors capture real-time data for simulations.
- Digital twin of systems: This class predicts and reflects the behavior of systems throughout their lifecycles, aiding tasks like real-time monitoring and predictive maintenance in various fields.

Digital twins in Industry 4.0 can be also categorized into types based on their characteristics:

- Plain gadget models: These models encompass current values obtained from sensors and expected values the gadget aims to achieve.
- Embedded digital twins (EDTs): EDTs actively participate in all operations involving their real twins, enabling smart decision-making through bidirectional connections between the physical and digital realms.
- Networked twins: Networking enhances connectivity and information exchange among integrated EDTs in smart manufacturing.

The most relevant feature is the level of integration which reveals the relation of the physical object with the digital instance. Digital twins are classified based on the level of integration of data between real and digital twins:

- Digital model: This virtual representation doesn't use automated data interchange between real and virtual objects. Data may be manually entered, and changes in one twin don't directly affect the other (Fig. 7.6).
- Digital shadow: Involves automatic unidirectional data interchange from real to virtual objects. Changes in the real object directly update the virtual twin (Fig. 7.7).

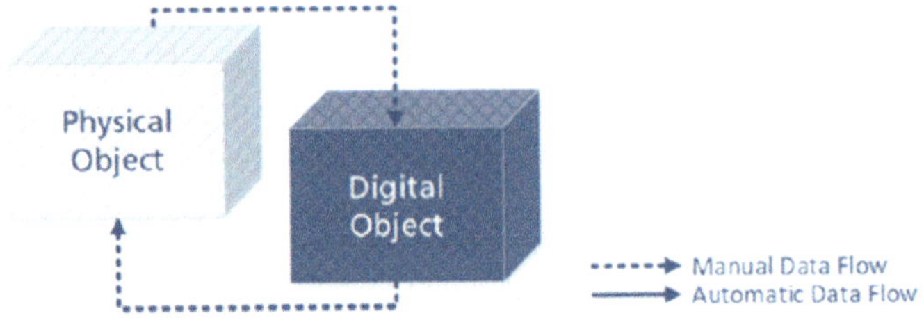

Fig. 7.6 No connection between model and real entity

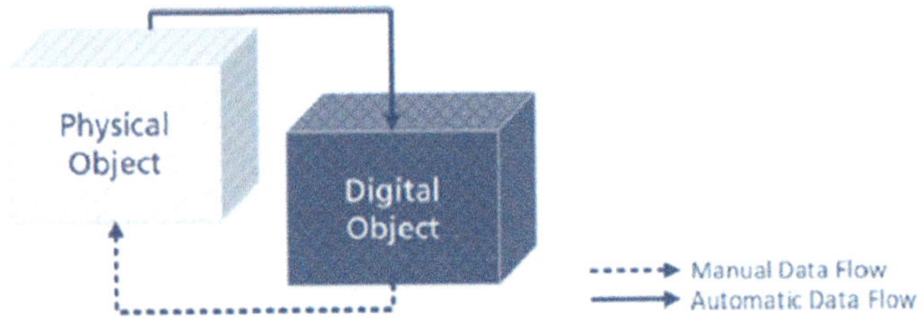

Fig. 7.7 One-way connection from real entity to reflection

- Digital twin: Features bidirectional data interchange between real and virtual objects, with changes in either twin directly affecting the other (Fig. 7.8).

A real digital twin exhibits three key characteristics (Fig. 7.8):

Real-time reflection: Digital twins maintain both physical and digital worlds, allowing synchronization through data exchange.
Communication and confluence: Digital twins involve communication and confluence within the physical world, between stored and current information, and between the physical and digital realms.
Self-evolution: Digital twins can refresh and modify real-time information, leading to positive changes in models and content as current information is compared with the physical world.

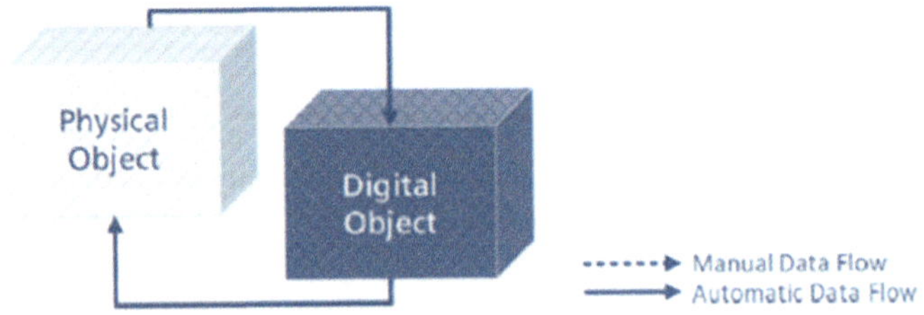

Fig. 7.8 Two-way connection and real twinning

Creating a Digital Twin Model

Creating a digital twin model is a complex endeavor, and there is no one-size-fits-all approach to building these virtual representations of real-world assets and systems. Different authors and practitioners employ different methods, methodologies, and modeling tools to develop these virtual counterparts (Makarov 2019). One of the most popular is the virtual prototyping powered by ALSTOM in energy and rail applications, as shown in Fig. 7.9, where physical models compensate for lack of knowledge extracted from data.

There are several variants to model a twin, but most comprise knowledge about the physics of the failure and the surrounding context of the asset:

(1) Systems Modeling Languages as a Basis for Digital Twins

Systems modeling languages play a crucial role in creating digital twins. These are graphical modeling languages designed to support the analysis, specification, design, verification, and validation of complex systems. They provide a structured framework for capturing essential aspects of systems, components, and objects. These aspects include:

- Structure, interrelation, and classification: Users can define the structural elements of a system, how they relate to each other, and their classification.

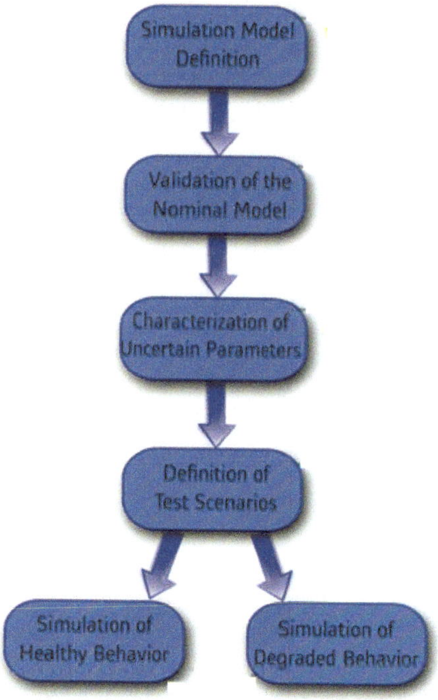

Fig. 7.9 Virtual prototyping process

- Behavior: They represent system behavior using functions, messages, and states, allowing a comprehensive understanding of how the system operates.
- Limitations: They permit the specification of physical and operational properties and constraints.
- Distribution: They help manage the distribution of elements, behavior, and limitations across a system.
- Requirements: They support the documentation of requirements and their relationships with other system conditions, design elements, and test cases.

2. Simulation as the digital twin foundation:

Simulation is a numerical method used to study complex systems by developing mathematical models of their elements and connecting these models into an informational representation. Hybrid models that combine AI with multiphysics simulations are becoming increasingly vital in the development of digital twins. These models harness AI's capabilities, including ML algorithms and data-driven insights, to augment the accuracy and predictive power of traditional simulations. One of the remarkable aspects of this integration is its ability to create synthetic data and compensate for the lack of real-world information (see Fig. 7.10). By incorporating AI into simulations, these hybrid models can analyze the complex interplay between various physical phenomena and forecast system behavior under diverse conditions. This integration leads to a more comprehensive understanding of intricate relationships among different variables, resulting in more precise and reliable digital twin models. Through AI-enhanced multiphysics simulations, organizations gain valuable insights into system performance, identify potential issues, and optimize operational strategies to enhance overall efficiency and resilience.

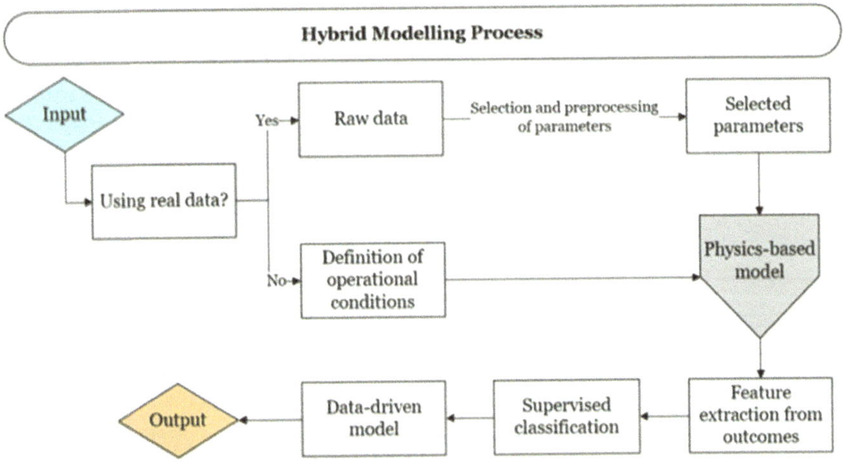

Fig. 7.10 Contribution of hybrid modelling to digital creation by means of simulation

When applying simulation to create digital twins, the following tasks must be addressed:

- Data acquisition: Gathering data from real-world objects is essential to create accurate virtual models that mirror physical objects.
- Software selection: It is necessary to decide whether to use universal simulation software or opt for custom development based on specific requirements.
- Effectiveness of digital twins: Digital twins are most effective in areas where formalized methods and mathematical models are integral.

Applications of Digital Twins

Digital twins play a pivotal role in enhancing the intelligence of operational systems across various industries. By maintaining an accurate and up-to-date representation of real-world operating assets, these virtual counterparts empower enterprises to exercise precise control and optimize both individual assets and the broader operational ecosystem. This representation encompasses not only the current state of assets but also their historical operational data. Digital twins offer a multitude of benefits, including optimization, automation, and predictive capabilities, and their utility extends to purposes beyond standard operations, such as virtual commissioning and the development of next-generation designs.

The key applications of digital twins span several domains:

Operations optimization: Digital twins excel in conducting what-if simulations that assess readiness and recommend adjustments. This capability enables organizations to optimize their operations, reduce risk, cut costs, and enhance overall efficiency. By running scenarios and evaluating potential changes, digital twins provide valuable insights into how to refine operational processes.

Predictive maintenance: Digital twins are invaluable in predicting the RUL of equipment and assets. By continuously monitoring and analyzing real-time data from these assets, digital twins can determine the optimal timing for maintenance or replacement. This proactive approach to maintenance helps organizations avoid unplanned downtime and costly repairs.

Anomaly detection: Operating in parallel with their real-world counterparts, digital twins are equipped to identify operational behavior that deviates from expected, simulated behavior. For instance, in the context of a petroleum company's offshore oil rigs that operate continuously, a digital twin can scrutinize sensor data to swiftly detect anomalies. This early detection is instrumental in preventing potential catastrophic damage or accidents.

Fault isolation: When anomalies are detected, digital twins can trigger simulations aimed at isolating the fault and identifying its root cause. This diagnostic capability empowers engineers or the system itself to take appropriate corrective actions promptly. By pinpointing the source of the issue, organizations can minimize downtime and ensure the safety and reliability of their assets.

In summary, digital twins serve as intelligent companions to real-world assets and systems, offering a range of capabilities that enhance operational efficiency, minimize risks, and optimize performance. Whether used for operations optimization, predictive maintenance, anomaly detection, or fault isolation, digital twins are instrumental in driving informed decision-making and ensuring the reliability and longevity of critical assets.

Some key sectors where digital twins are making a significant impact are:

- Manufacturing: Digital twins are revolutionizing the manufacturing industry by optimizing product design, production processes, and maintenance procedures. This optimization leads to reduced throughput times and enhanced operational efficiency, ultimately resulting in cost savings.
- Automobile: In the automotive sector, digital twins create virtual models of connected vehicles, capturing comprehensive behavioral and operational data. This data analysis aids in evaluating overall vehicle performance as well as individual connected features. Digital twins also enable personalized customer service, enhancing the automotive user experience.
- Retail: Digital twin technology is enhancing the retail industry by offering virtual representations of customers and allowing the modelling of fashion items on these digital avatars. This capability improves customer experiences by enabling personalized shopping recommendations. Digital twins are also used to optimize store planning, enhance security measures, and manage energy resources efficiently.
- Healthcare: Combining digital twins with IoT data has far-reaching applications in healthcare. These applications range from cost-saving measures to patient monitoring, preventative maintenance of medical equipment, and personalized healthcare solutions. Digital twins facilitate better patient outcomes and resource management in healthcare settings.
- Smart Cities: Digital twins, coupled with IoT data, play a crucial role in the development of smart cities. They contribute to economic growth, efficient resource management, reduced environmental impact, and an improved quality of life for residents. City planners and policymakers use digital twin models to access data from various sensor networks and intelligent systems, enabling more informed decision-making for urban development.
- Industrial IoT: In industrial settings, digital twins empower firms to monitor, track, and control industrial systems digitally. Beyond operational data, digital twins capture environmental data, including location, configurations, and financial models. These data enable the prediction of future operations and anomalies, enhancing operational efficiency and cost-effectiveness.

The digital twin concept has a transformative impact on various industries, offering opportunities for optimization, innovation, and data-driven decision-making. Its application extends to manufacturing, automotive, retail, healthcare, smart cities, and IIoT, where digital twins enhance performance, customer experiences, and resource management, while driving cost savings and operational efficiency.

Applications of Digital Twins Through the Product Lifecycle

The application of digital twins is a multifaceted concept that spans various stages of the product lifecycle and industrial processes.

- Product design and optimization: Digital twins are increasingly integrated into the product design stages, offering a quantitative tool for efficient and optimal decision-making. Data from previous product generations are amalgamated to form a comprehensive digital twin, facilitating knowledge transfer and enhancing the early stages of new product development. This approach leverages data from digital twins of past product designs to analyze and optimize new designs, streamlining the design process.
- Production and manufacturing: Digital twins have a significant presence in production systems. They are used to simulate production processes, predict outcomes, optimize operations, correct deviations, and evaluate system performance. By modeling manufacturing steps and entire machine tools, digital twins help determine the effects of tool behavior and process parameters, leading to optimized tool geometries and enhanced product quality. In additive manufacturing, digital twins are employed to evaluate 3D printed metallic components, reducing trial and error tests and shortening the design-to-production timeline. Complex production systems, characterized by interconnected manufacturing, quality control, and logistics processes, also benefit from digital twins. These systems involve stochastic and dynamic processes with non-linear dependencies that are challenging to address analytically.
- Optimization: Simulation models of digital twins are eventually used in combination with optimization programs to achieve various objectives, including selective part assembly, robust production scheduling, and the prediction of countermeasures in response to disturbances.
- Maintenance: Maintenance and refurbishment play key roles in shaping the behavior of assets, introducing new components, and even involving suppliers outside the original equipment manufacturer (OEM) supply chain. Third-party maintenance providers, with the appropriate service-level agreements, can modify assets independently of the OEM. This scenario imposes an obligation on asset owners or operators to maintain an up-to-date digital twin that accurately represents the asset's as-maintained state. Even if operators are not directly connected to the manufacturing supply chain, they must ensure the digital twin remains relevant. This requirement extends to facilitating the seamless handover of digital twin data from the manufacturing process to the operating process owner (Bächle and Gregorzik, 2019).

Despite the evident benefits, current approaches to digital twins often operate within distinct and separate disciplines. This siloed approach can lead to missed opportunities. Product design and specification may occur without considering more efficient production possibilities, and highly precise production processes may not take advantage of previously acquired product knowledge and the interactions of individual features.

In essence, digital twins offer a versatile set of tools that can revolutionize product design, manufacturing, and production systems. Their ability to simulate and optimize processes, predict outcomes, and facilitate human-robot collaboration holds immense potential for industries seeking to enhance efficiency, reduce costs, and accelerate development timelines. However, to realize these benefits, there is a need for greater integration and collaboration across disciplines to ensure that knowledge and insights from digital twins are leveraged holistically throughout the product lifecycle.

Digital Twins and Predictive Maintenance

A digital twin is a dynamic digital replica of a physical entity, bridging the gap between the physical and virtual worlds. It leverages IoT, AI, ML, and software analytics to create simulation models that continuously adapt to changes in their physical counterparts. Maintenance analytics is a crucial component within the context of digital twins and Industry 4.0. By integrating data-driven insights and analytics, maintenance processes can be optimized for efficiency and cost-effectiveness. Maintenance analytics leverages historical and real-time data to identify patterns, anticipate equipment failures, and schedule preventive maintenance tasks. This proactive approach ensures potential issues are addressed before they result in costly downtime or disruptions to production. With the integration of maintenance analytics, organizations can make informed decisions based on data-driven predictions, leading to improved asset performance, extended equipment lifespan, and overall operational resilience (Fig. 7.11).

Essentially, a digital twin evolves and updates itself based on multiple data sources, offering real-time insights into its present and future states.

Industry 4.0 has ushered in a strategic shift from reactive to predictive maintenance. Predictive maintenance assesses equipment conditions through periodic or continuous monitoring. The objective is to perform maintenance at the most cost-effective moment, just before equipment performance falls below a certain threshold. Digital twins have the potential to elevate predictive maintenance to the next level.

Analytical solutions for predictive maintenance empower organizations to proactively prevent unforeseen events and monitor asset conditions or entire production processes. When combined with the capability to simulate behavior, digital twins enable companies to optimize operations comprehensively and efficiently. They also facilitate testing of production developments and planned investments. Collaboration between predictive maintenance platforms and digital twins becomes crucial.

By simulating asset behavior and maintenance scenarios, digital twins inform decision-makers about critical maintenance Key Performance Indicators (KPIs) such as cost, downtime, RUL, end of life (EoL), and mean time between failures (MTBF). These simulations empower enterprises to plan future maintenance, enhance preventive and condition-based maintenance processes, and minimize unscheduled downtime (ReliaSol 2021).

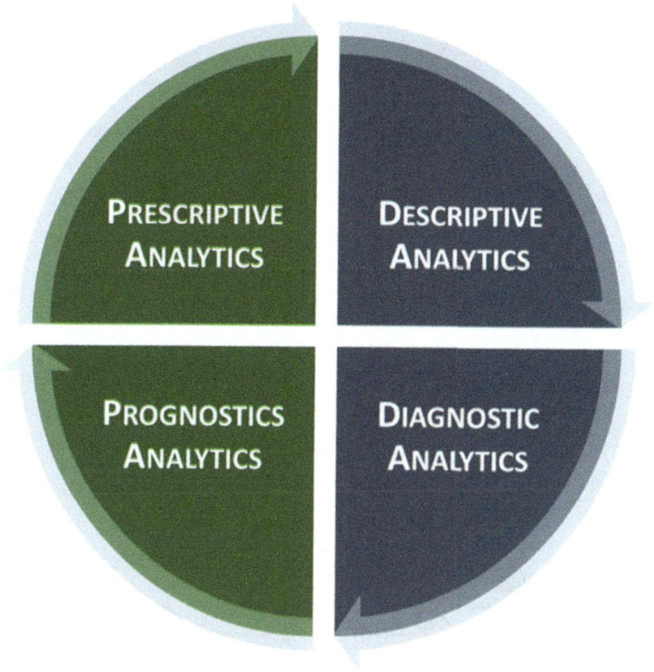

Fig. 7.11 Maturity stages in maintenance analytics

How Are Digital Twins Used in Maintenance?

Digital twins find significant utility in maintenance in the following areas:

- Digital simulation: Digital twins provide essential data for realistic asset behavior and maintenance simulations. These simulations consider risk factors, failure modes, operational scenarios, and system configurations. They yield maintenance-related KPIs like cost, downtime, RUL, EoL, and MTBF. Simulations support predictive maintenance planning and improve preventive and condition-based maintenance processes, minimizing unplanned downtime. Indeed, reliability, availability, maintainability, and safety (RAMS) knowledge during design is crucial to cover all the ways an asset might fail and therefore increase the digital twin detectability or predictability of such failure modes as depicted in the flowchart in Fig. 7.12.
- What-if analysis: Organizations leverage digital twins to simulate various maintenance scenarios (Fig. 7.13), aiding in the selection of the most effective strategy. These analyses contribute to long-term planning decisions, such as choosing between predictive and preventive maintenance strategies, and short-term choices like asset replacement.
- Maintenance system configuration: Digital twins synchronize with the status of their physical counterparts. Changes in the asset's status reflect in the digital twin

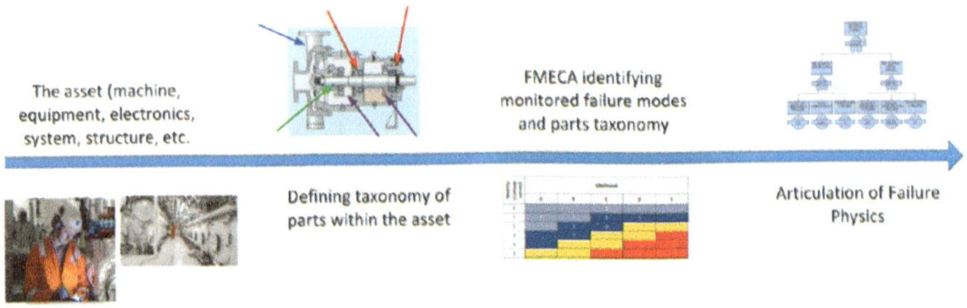

Fig. 7.12 Digital twin creation based on design information and RAMS parameters

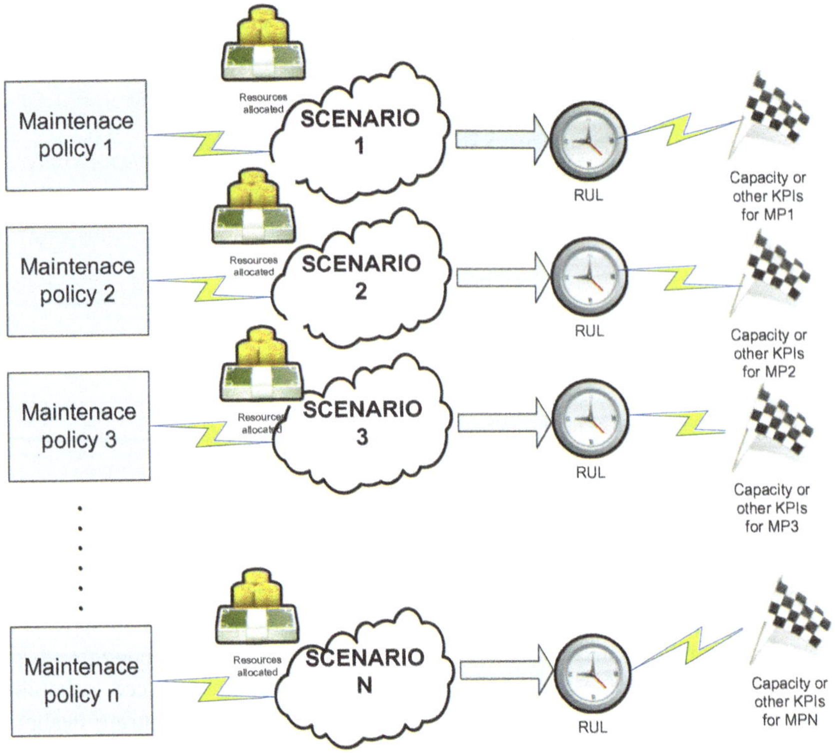

Fig. 7.13 Prescriptive analytics for maintenance performed in a digital twin for the optimal decision support system (DSS)

and vice versa. This synchronization allows digital twins to configure asset operation and related physical systems. Systems can adjust their physical components based on information and commands from their digital counterparts.

- Innovation: Digital twins serve as innovation catalysts in maintenance. They facilitate the testing, validation, and evaluation of innovative maintenance concepts without disrupting operations.

In fact, digital twins are transformative tools in maintenance, enabling data-driven decision-making, scenario analysis, and improved asset performance throughout the lifecycle.

Digital Twin Implementation Considerations

Digital twin technology represents an advancement in numerous industries, offering a holistic approach to enhancing operational efficiency and informed decision-making. This innovation allows organizations to create digital replicas of physical assets, enabling profound analysis and real-time monitoring. However, successful implementation requires careful consideration of multiple factors, including reference models, regulatory compliance, implementation phases, and organizational approaches.

Regulatory requirements: Digital twin technology serves as an invaluable tool for organizations aiming to conform to regulatory requirements. Particularly in industries subject to strict environmental regulations, such as automotive, marine, and aerospace, digital twins enable engineers to redesign components, significantly reducing emissions and helping organizations avoid regulatory fines and expenses (Altair 2019).

Timeline considerations: The timeline for implementing digital twin technology varies significantly based on factors such as the type of asset, accuracy requirements, feasibility, cost considerations, and technology readiness. At its core, a basic digital twin implementation necessitates:

- Edge capabilities, which encompass observing key aspects of the asset's real-time state and behavior. This typically involves deploying sensors with associated edge processing capabilities and enhancing data quality through processes like calibration, filtering, and time synchronization.
- Digital twin core runtime, which utilizes the data stream from the edge to create a (near) real-time digital representation of the asset's state.

The application layer integrates with the digital twin's data streams, becoming an integral component of various business processes, including user applications for monitoring and control, legacy applications for maintenance and asset management, and data analytics and ML stacks for pattern recognition and decision support (Erikstad 2017).

Practical considerations in implementation: A well-structured approach is crucial to the success of digital twin implementation (Roundy 2020). This organization involves four essential steps:

- Involving the entire product value chain: Collaboration across the product value chain is vital. Different departments within an organization face distinct business

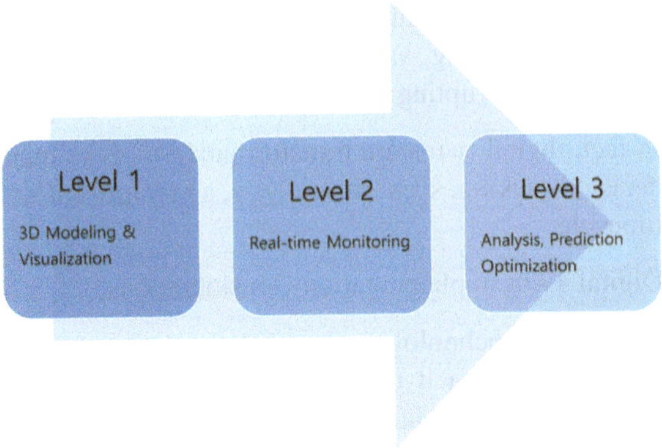

Fig. 7.14 Evolution from 3D models to predictive engines

challenges in their daily operations. A digital twin offers solutions to issues such as cross-functional collaboration, data-driven decision-making, and supply chain coordination. Gathering insights and inputs from stakeholders at all levels ensures a more efficient digital twin design.

- Establishing well-documented practices: Employing standardized and well-documented design practices enhances transparency and simplifies collaborative work. This approach fosters the communication of ideas across departments and regions, allowing multiple users to build or modify digital twin models without disrupting existing components.
- Incorporating data from multiple sources: A rich dataset from various sources, both internal and external, is fundamental for creating realistic and insightful simulations. While 3D modeling and geometry are sufficient for representing how parts fit together and how a product functions, predicting faults and errors requires extensive data and advanced analytics. Figure 7.14 shows the evolution of digital twins from 3d representations to analytic engines.
- Ensuring long access lifecycles: Avoiding vendor lock-in is crucial when implementing digital twins using proprietary design software (Roundy 2020). Assets with long lifecycles, such as buildings and industrial machinery, often outlast the software used to design them. To mitigate this risk, IT architects and digital twin owners should establish terms with software vendors to ensure ongoing data compatibility and avoid dependency on a single supplier.

In conclusion, digital twin technology offers immense potential when organizations pay close attention to reference models, regulatory requirements, structured implementation phases, industrial applications, and the broad range of benefits. The success of digital twin implementations is closely tied to embracing an inclusive approach, fostering standardized practices, leveraging diverse data sources, and ensuring long-term sustainability in the deployment of this transformative technology.

Cost of digital twin implementation: When implementing digital twin technology, various factors contribute to the overall costs, and the expected returns on investment (RoI) largely depend on the specific application and the scale of the asset systems.

The cost of implementing and starting up digital twins can vary significantly based on asset type, size, complexity, and the level of detail required by the client (Lengthorn 2021). Not all sectors have a quick payback in terms of ROI. The benefits and RoI of digital twins are particularly pronounced in the maintenance sector. The application of DTs in maintenance yields several positive impacts:

- Insights into asset management: Digital twins provide insights into asset management processes, enabling the optimization of maintenance strategies by identifying non-obvious failure or degradation patterns.
- Optimal maintenance decisions: Simulations enabled by digital twins facilitate optimal maintenance decisions, leading to improved Overall Equipment Efficiency (OEE) and better RoI.
- Automation and cost-effectiveness: Digital twins increase the automation and cost-effectiveness of maintenance processes, enhancing their flexibility.
- Transition to predictive maintenance: Digital twins aid in the transition from traditional maintenance approaches to more effective ones, like predictive maintenance, with minimal disruption to operations (Edge4industry 2018).

However, the design and construction of digital twins for maintenance applications remain costly and complex. To harness the benefits, various aspects must be considered:

- Understanding assets' physical properties, including electrical and mechanical specifications.
- Identifying failure modes, their criticality, and degradation patterns.
- Incorporating statistical information, such as failure probabilities and distribution functions.
- Aligning digital twins with maintenance and business goals, including cost targets, spare parts inventory, OEE, and risk management (Edge4industry 2018).

A phased approach, starting with simpler models and gradually incorporating more sophistication, is a practical way to implement digital twins while minimizing risks and gaining confidence in their use maybe less costly and a right approach for quick wins.

Digital twins have a significant impact on maintenance indicators by optimizing processes, improving decision-making, and increasing automation. They can lead to substantial cost savings, particularly in maintenance costs, by identifying and addressing issues proactively (Edge4industry 2018). Furthermore, digital twins have a substantial impact on the income statement of organizations. By integrating technologies like artificial intelligence, machine learning, and software analytics with real-time data, digital twins create simulation models that optimize development cycles, anticipate downtime, and enable real-time performance assessment (TWI 2021). However the ROI should be considered before deciding the adoption of such technology and the timing to design and deploy.

Conclusion: Advantages of Digital Twin Technology for Maintenance and Ongoing Issues

1. Advantages of digital twins in maintenance

Many companies across diverse industries are actively investing in digital twin technology, offering digital twin software solutions, or applying digital twins within their own operations (Sharma 2020). The adoption of digital twin technology has compelling benefits for enterprises, including:

- Continuous asset tracking: The ability to monitor assets, components, and processes in real-time.
- Efficient problem understanding: Quick identification and understanding of issues as they arise.
- Enhanced product and operation improvement: Opportunities to refine products, processes, and services based on real-time insights.
- Facilitation of innovation: Reduced risks associated with high-cost investments.
- Advanced planning through simulations: Improved planning and decision-making through the use of simulations.
- Effective problem tracing: The ability to pinpoint and address issues that traditional methodologies may miss.
- Predictive maintenance: Anticipating failures in terms of their type and timing, enabling proactive maintenance (ReliaSol 2021).

The latter point is especially important. Many assets, especially complex and long-lasting ones like aircraft, ships, locomotives, and wind turbines, undergo substantial changes throughout their operational lifespans. This necessitates efficient maintenance decision-making to prevent unscheduled maintenance, as it can lead to increased costs and operational delays. Predictive analysis has emerged as a key strategy for improving reliability and reducing unscheduled maintenance. Organizations are increasingly turning to predictive analysis to anticipate potential failures before they occur, addressing long-standing issues related to asset failures.

The digital twin plays a pivotal role in enabling effective predictive analysis. It serves as an exact replica of a physical asset, offering the essential context required for accurate predictions. This context encompasses the entire history of an asset, capturing its configuration and managing changes over time. The digital twin integrates data from various sources, including computer-aided design (CAD), simulation models, IoT data, time series data, and maintenance records, to provide a comprehensive and detailed picture of an asset's condition.

However, it's crucial to distinguish between digital models and digital twins. While there is a growing trend to use simulation or CAD models as digital twins, this approach can be problematic. These models may not accurately reflect the final as-built configuration of an asset, as changes often occur during manufacturing, modifications, and defect rectification. As assets undergo maintenance and upgrades over time, they may deviate significantly from the original models.

To maintain the effectiveness of predictive maintenance, it is important to continuously update the digital twin to reflect significant changes in an asset's configuration. This includes capturing alterations made during maintenance, such as component replacements. This constant updating ensures the digital twin configuration provides the necessary context for accurate predictive analytics. For example, it allows different maintenance approaches based on specific asset configurations, even if two assets have logged the same number of operating hours.

Predictive maintenance, powered by context-rich digital twins, relies on real-time data from IoT sensors. These data are sent to the digital twin configuration, where they are analyzed against OEM specifications. Multi-physics simulation models are then applied to interpret the data and predict potential component failures proactively.

Challenges and Barriers to the Adoption of Digital Twin Technology

Despite the promise of digital twins, their implementation can be challenging. Several common errors and pitfalls should be avoided to ensure success. These include repurposing a digital twin platform for different applications, attempting to implement digital twins across an entire production line or facility too quickly, neglecting data quality control, overlooking the importance of device communication standards for IoT devices, and failing to secure buy-in from users across the product value chain. Addressing these challenges and avoiding common pitfalls is essential to maximize the effectiveness and value of digital twin implementations across various industries.

- Digital twins focus on providing insights into physical systems. This limits their applicability in certain fields that require a more holistic understanding. For example, when used in urban planning, digital twins cannot address underlying sociopolitical issues, such as social inequality or housing crises. Thus, their use may not directly impact broader societal challenges (Kshetri 2021).
- Digital twin adoption may be hampered in developing economies, primarily due to the computational power required to create high-fidelity models, which can often exceed the available resources (Kshetri 2021).
- Cost is another issue, especially for projects with short lifespans. Implementing and maintaining digital twins can be prohibitively expensive, potentially undermining their viability (Sharma 2020).
- The intricate nature of digital twin technology further complicates matters. It demands seamless integration of various components, real-time tools, algorithms, and vast amounts of Big Data, a process that can be time-consuming and resource-intensive (Sharma 2020).

  Digital twins require continuous updates to remain aligned with advancements in related technologies and remain fit-for-purpose throughout their lifecycle (Sharma 2020). Correctly designing a DT to carry out its intended purpose and evaluating its performance are non-trivial tasks.

- Digital twins need an extensive and reliable data supply to function effectively. They produce copious amounts of data of various types, and users must be able to swiftly access and extract meaningful insights from this data. System knowledge often proves incomplete, inconsistent, or erroneous, posing challenges to data quality. Furthermore, issues related to sensitive data, such as privacy concerns and the protection of business secrets, can complicate digital twin development. Differing stakeholder perspectives on data quality, based on their unique purposes, add further complexity to the challenge. Accountability and transparency in data usage are essential to foster user confidence in the results (Pileggi 2021).
- In the realm of model fidelity, the challenge lies in determining which features of the system are most salient and relevant. Striking the right balance between too much detail, which can be costly and complicated, and too little detail, which may be insufficient, requires careful consideration. As the purpose of the digital twin evolves, the model, data infrastructure, and applications must be continually evaluated and adapted (Pileggi 2021).
- Maintaining the reliable operation of digital twins is another technical challenge. Digital twins are designed to be used throughout the lifecycle of a real-world object or system, which entails managing a complex blend of software, hardware, measurements, and simulations. This complexity escalates when various stakeholders from different organizations are involved concurrently. Maintenance, encompassing software upgrades, hardware component changes, and model adjustments, is crucial to ensuring that a DT continues to deliver value efficiently (Pileggi 2021).

  Effective use of digital twins necessitates the careful allocation of computational resources, often distributed across private and public clouds, vendor platforms, and high-performance computing resources. Security considerations must be integrated, and the system must be designed to prevent computational overload (Pileggi 2021).
- The absence of standardized definitions, common language, and established best practices in the industry poses a significant technological challenge. This lack of standardization can make it difficult to identify and address specific requirements for digital twin implementations.
- Dgital twins often require data related to product lifecycle management, which may come from a company's suppliers and even their suppliers' suppliers. Obtaining access to data on products and processes outside an organization can present challenges, including issues of data sharing and integration (Lawton 2020). To foster the development and widespread acceptance of digital twins, the industry needs to create standards and best practices. Establishing common definitions, language, and guidelines can facilitate smoother implementation and interoperability across various digital twin systems. These standards will play a crucial role in realizing the full potential of digital twins across diverse sectors.

In summary, the practical application of digital twins faces myriad challenges encompassing technical, knowledge-based, and organizational aspects. Overcoming these barriers is essential to fully leverage the potential of digital twins while recognizing their inherent limitations.

References

Altair (2019) Why all the buzz about digital twins? https://www.altair.com/newsroom/articles/why-all-the-buzz-about-digital-twins/

Bächle K, Gregorzik S (2019) Digital twins in industrial applications—requirements to a comprehensive data model. CONTACT Software GmbH. IIC J Innov

Bauer W, Schlund S, Marrenbach D, Ganschar O (2015) Industrie 4.0 – Volkswirtschaftliches Potenzial für Deutschland. Bitkom und Fraunhofer IAO 2015

Dittrich M-A, Schleich B, Clausemeyer T, Demgrave R, Erkonyuncu J, Häfner B, Lange J, Scheidel W, Wuest T, Plakhotnik D (2019) Shifting value stream patterns along the product lifecycle with digital twins. Towards shifted production value stream patterns through inference of data, models, and technology. 7th CIRP global web conference

Edge4industry (2018) Digital twins: what are they and how are they shaking up enterprise maintenance? https://www.edge4industry.eu/2018/06/19/digital-twins-what-are-they-and-how-are-they-shaking-up-enterprise-maintenance/. Accessed 25 Aug 2021

Happiest Minds (2021). Digital Twins. https://www.happiestminds.com/insights/digital-twins/. Accessed 24 Aug 2021

Juarez MG, Botti VJ, Giret AS (2021) Digital twins: review and challenges. J Comput Inf Sci Eng 21:030802–1

Lawton G (2020) Early days for digital twins in the supply chain. TechTarget

Lengthorn P (2021) How Grieves M., 2014. Much to build a Digital Twin?. https://www.consultengsurvivor.com/cost-to-build-a-digital-twin. Accessed 22 Aug 2021

Makarov VV, Frolov Ye B, Parshina IS, Ushakova MV (2019) The design concept of digital twin. In: 2019 twelfth international conference "Management of large-scale system development" (MLSD). IEEE, Moscow, Russia. https://doi.org/10.1109/MLSD.2019.8911091

Nir Kshetri N (2021) The economics of digital twins. University of North Carolina at Greensboro. Digital Object Identifier https://doi.org/10.1109/MC.2021.3055683. Accessed 9 Apr 2021

Ove Erikstad S (2017) Merging physics, big data analytics and simulation for the next-generation digital twins. Norwegian University of Science and Technology. HIPER 2017, High-Performance Marine Vehicles, Zevenwacht, South-Africa, 11–13 September 2017

Pileggi P (2021) Overcoming 9 Digital Twin barriers for manufacturing SMEs. Position Paper. Change2Twin

Rasor R, Göllner D, Bernijazov R, Kaiser L, Dumitrescu R (2021) Towards collaborative life cycle specification of digital twins in manufacturing value chains. In: 28th CIRP conference on life cycle engineering. Proc CIRP 98:229–234

ReliaSol (2021) Digital twins & predictive maintenance—solutions that are changing the industry. https://reliasol.ai/digital-twin-predictive-maintenance-technologies-that-change-the-industry/. Accessed 22 Aug 2021

Roundy R (2020) Best practices for digital twin implementation. https://dzone.com/articles/best-practices-for-digital-twin-implementation-1. Accessed 25 Aug 2021

Sharma A, Kosasih E, Zhang J, Brintrup A, Calinescu A (2020) Digital twins: state of the art theory and practice, challenges, and open research questions

Singh M, Evert FE, Hinchy EP, Qiao Y, Murray N, Devine D (2021) Digital twin: origin to future. Appl Syst Innov 2021(4):36. https://doi.org/10.3390/asi4020036

TWI (2021) What is digital twin technology and how does it work? The Welding Institute. https://www.twi-global.com/technical-knowledge/faqs/what-is-digital-twin. Accessed 22 Aug 2021

Van der Valk H, Haße H, Möller F, Arbter M, Henning J-L, Otto B (2020) A taxonomy of digital twins. In: Americas conference on information systems

Chapter 8
Digital Twin for RAMS

Bhupesh K. Lad, Ram S. Mohril, Ishika Budhiraja, and Joydeep Majumdar

Introduction

Industry 4.0 is changing the way systems are conceived, built, operated, maintained, reused, and discarded. This revolution is focused on developing digitally assisted intelligent assets that will function as part of a Cyber-Physical System (CPS). Digital Twins (DTs)are the foundational elements of any CPS. With features like real-time synchronization, secure communication, and enhanced capabilities for analytics and simulations, DT is poised to revolutionize the reliability engineering and maintenance management domain to a significant extent. With such integration of DT, the asset will truly emerge as an intelligent asset. This intelligent asset will not only be self-aware but also situationally aware. DT is expected to manifest and embed itself in the various RAMS practices and disrupt their functioning throughout the life of an asset. At the design and testing phase of the asset, DT is expected to leverage the capability of the simulations to severely influence the conventional methodologies of reliability testing. These DT-driven analytical models are largely expected to disrupt the whole utilization phase by leveraging the power of AR-VR-XR in maintenance assistance; and with the capability to communicate, making use of technologies like transfer or collaborated learning into the parallelly emerging applications of predictive maintenance. With the capability of DT to make the decision at distributed levels and in a decentralized manner, the assets and, further, the organizations will be able to make critical decision making in near real-time, making the systems much more resilient. The evolving DT architectures facilitate the systems to make secure use of several third-party applications for executing important processes like FTA, FMECA, RCM, etc. With the smart integration of human reliability models, these DT-enabled assets will take human–machine interactions to the next level.

B. K. Lad (✉) · R. S. Mohril · I. Budhiraja · J. Majumdar
Department of Mechanical Engineering, Indian Institute of Technology Indore, Indore, Madhya Pradesh, India
e-mail: bklad@iiti.ac.in

P. V. Varde et al. (eds.), *Advances in Risk-Informed Technologies*, Risk, Reliability and Safety Engineering, https://doi.org/10.1007/978-981-99-9122-8_8

This paper initially presents a birds-eye view of the potential disruption that DT would cause to existing RAMS and later presents a case study of selective maintenance to highlight the potential of DT in RAMS.

Digital Twins and RAMS

The term 'Digital Twin' is defined by ISO23247 as 'a suitable digital representation of some realized thing or process with a means to enable convergence between the realized instance and digital instance at a suitable rate of synchronization' (ISO Standard 2021). DT technology is expected to improve every stage of the product life cycle (Khalyasmaa et al. 2023). Similarly, RAMS is important throughout the product life cycle. Working with next-generation intelligent systems necessitates tailoring the RAMS analysis with DT.

Creating DT for RAMS

An important aspect of DT creation is its architecture. Architecture facilitates the overall working of DT, including data collection, storage synchronization, modeling, analysis, and decision making. In addition, it should facilitate communication, access and control, data security, and overall management of DT. ISO 23247 (ISO Standard 2021) provides a reference architecture. It covers different domains and various modules supporting its functionality. Figure 8.1 shows various domains used in ISO 23247. This reference architecture may be used to build customized architecture to suit desired RAMS applications. In the case of RAMS, the observable manufacturing domain could be the asset i.e., product, system, or component for which the analysis is required. Data collection and control domain facilitate static and dynamic data collection related to the asset. Static data may include three-dimensional scans, equipment make, model, bills of materials, costs, maintenance plans, and warranties (EP Editorial Staff 2022). Dynamic data may include current age and condition indicators such as pressure, temperature, flow, electrical current, and vibration (EP Editorial Staff 2022). Some of these data are generated prior to the usage phase. In the design phase, for example, the system configuration and failure probability distribution parameters of the components are finalized. As a result, DT must be built from the concept phase of the product life cycle and evolve throughout the product's existence. Managing such evolution of DT provides another important research area. Data collected from diverse sources must be synchronized with the physical system. This feature is included in the DT core domain in reference architecture. In addition, all analytical models and algorithms will reside in this domain. For example, an algorithm to estimate reliability parameters may be included in this domain in a separate micro module. Similarly, a maintenance optimization algorithm can be incorporated.

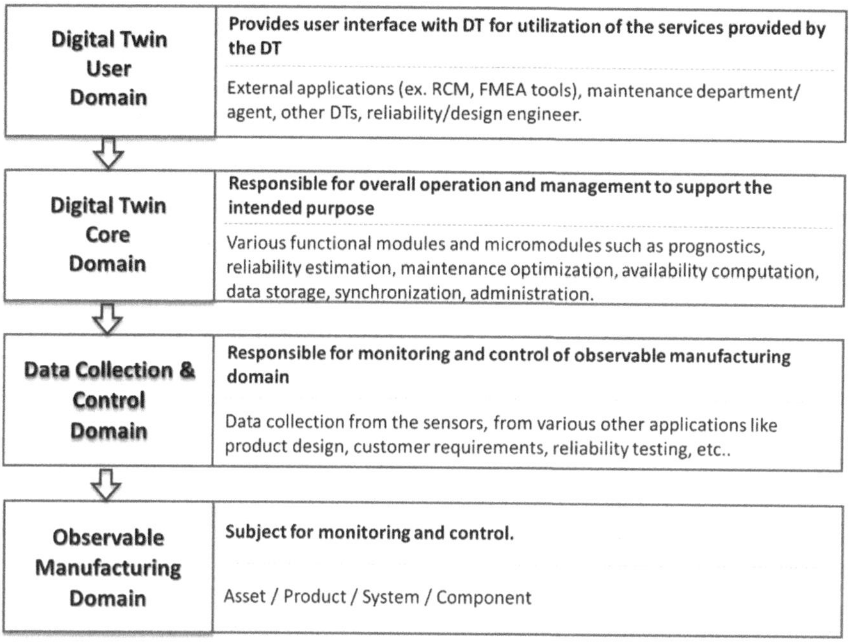

Fig. 8.1 Domains of DT defined by ISO 23247

In many cases, DT may want to use existing tools; for example, commercially available reliability software for reliability estimation may be used. In such cases, external entities will form part of the DT user domain. DT core domain must facilitate access and communication with such external applications.

DT Characteristics and RAMS

Some of the typical characteristics of DT include awareness (Jeon et al. 2020; Hribernik et al. 2021), communication (Wu et al. 2021), learning (Jay et al. 2019), and autonomous decision making (Hribernik et al. 2021). DT of an asset should have a complete idea about its failure and repair characteristics under various operating and maintenance conditions, possible failure modes, baseline statistics, and the current health of its physical counterpart. This can be related to self-awareness. In addition, DT should be situationally aware, i.e., it should know its operating environment, such as load, humidity, temperature, duration, etc. DT may be required to share or exchange data/information with other DT or any third-party applications. It relates to the communication characteristics of the DT. In the case of RAMS, data/information exchange may include failure and repair parameters, its current health, and decisions taken at various levels. In addition, DT must learn from its own data and other relevant DTsin the network. For example, DT may identify similarities or dissimilarities with other assets in the network and extract appropriate learning to tune its own algorithms. Such transfer learning-based algorithms will help in improving the accuracy of prognostics and health management. A DT of an asset should be able to make various decisions with or without humans in the loop. For example, the machine must be able to suggest its own maintenance plan or perform RCM/FMEA, etc., for the same DT will comprehend the data and learn and react. It may be required to communicate with external applications like RCM tools or FMEA tools. Another example of autonomous decision making is where DT helps in learning from the environment and modifying operating parameters to successfully complete a mission.

Application of DT in RAMS

DT will cause disruption to the RAMS field. A brief overview of some of the potential areas is given hereunder.

Reliability design: DT will revolutionize reliability design. The future of reliability design is generative reliability design. Initially, generative design was used as a 3D CAD feature that uses AI to produce optimal designs from a set of system design requirements. The purpose of generative design is to use AI algorithms to generate and evaluate a variety of alternative design iterations based on user input. DT is a promising technology to deliver generative design. Performance criteria (such as mission reliability and availability), cost, weight, slope of the reliability growth curve, potential failure modes under specific operating environments, and available alternatives for components/subassemblies can all be factored into this design process to produce the best possible results by situationally aware DT of the system. DT will investigate the design space using machine learning algorithms based on data from previous products as well as data acquired through physical sensors. The DT's learning capabilities will further affect the generative reliability design process by driving 'closed-loop' procedures in which performance improves with each design iteration or system reconfiguration.

Reliability Prediction: One of the important areas in RAMS is the reliability prediction based on the field data. However, reliability engineers suffer due to a lack of accurate data from the field. The existing body of literature indicates that reliability prediction should involve a sophisticated integration of various domain-specific factors. These factors include environmental conditions, terrains (specifically for mobility assets), the impact of human intervention in operation and maintenance, operational stresses, utilization of reconditioned spares, as well as structural and economic dependencies. DT of assets will provide accurate information about failure modes, operating environment, and stress conditions for more accurate reliability prediction. However, researchers need to develop models to use such information for reliability estimation and integrate them with DT.

Maintenance Management: Areas where DT plays vital roles in maintenance management include maintenance optimization, predictive maintenance, and remote maintenance assistance. Conventional maintenance optimization problems rely on centralized data collection and processing. Such maintenance optimization problems for real-life systems are very complex and demand high computational time. As s result, simplified assumptions are made, which make the problem far from real-world systems. DT provides solutions to such challenges by allowing the use of distributed intelligence and computation power to model and solve the problem in a decentralized manner. This not only can reduce computation time but also allow better local-level exploration, thereby making the solution closer to the actual scenario (Upasani et al. 2017). In addition, computation efficiency makes such approaches suitable for dynamic decision-making requirements of next-generation CPS. The present paper discusses a DT-driven decentralized approach in detail in the next section.

Another area where DT helps the maintenance engineer is predictive maintenance. Prognostic approaches are widely used for the same. Though prognostics is an older term compared to DT, in CPS, the utilization of DT allows for prognostics to be executed at the edge in a distributed manner. The use of transfer learning further improves the potential of predictive maintenance (Palau 2019). Transfer learning involves the collaboration of DTs through three distinct mechanisms: the sharing of sensory information, the sharing of the consequences of their decision making, and the sharing of decision rules.

Apart from these, one can create an immersive environment for maintenance training and remote maintenance by adding additional layers of Augmented Reality (AR), Virtual Reality (VR), and Mixed Reality (MR) in DT (Eswaran and Bahubalendruni 2022).In order to enable this, at the design stage, designers must incorporate interactive 3D models as the metadata of the product in its design stage. AR/VR tools/applications can be used as the user domain for asset DT for creating required maintenance training.

Reliability Centered Maintenance (RCM): RCM is used as a process to identify an asset's maintenance strategy. It mainly relies on operating and maintenance data kept in Excel spreadsheets and other heterogeneous maintenance management systems. These semi-structured or unstructured data make it very difficult to completely automate the RCM process. Thus, it becomes unsuitable for any CPS. DT powered with ontology-based data collection will not only help in automating RCM but also support various other RAMS methodologies like FMEA/FMECA. For example, work done by Mohamed-Hedi et al. (2011) defines Industrial Maintenance Management Ontology (IMMO). Similarly, Jimenez et al. (2023) defined an ontology model for maintenance strategy selection. Building RCM ontology based on such works and integrating it with DT architecture provides scope for further development in this area.

In order to gain deeper insight into how a DT based approach be developed and used with DT, the following section provides a case study for DT based decentralized selective maintenance optimization. It also highlights the benefits of the proposed approach over the conventional approach and its suitability for next-generation CPS.

DT Based Decentralized Approach for Selective Maintenance Optimization

This section describes a case study that shows how DTs can be used in decentralized decision making to optimize selective maintenance. For mission-critical assets operating in the mission mode, the system components must undergo maintenance during the planned breaks to ensure they execute their next task effectively. Few components can be maintained during scheduled breaks due to short break duration and limited maintenance resources. Therefore, the best subset of system components and their maintenance processes must be chosen to achieve the minimum reliability level needed for the next operation. The high computation time involved in solving such a combinatorial selective maintenance optimization problem is undesirable because it directly consumes actual maintenance time from the limited interval period.

We hypothesized that by leveraging the features of DT in CPS, such selective maintenance problems can be solved in a decentralized manner in significantly less computation time without compromising the solution quality. We developed a methodology to solve the selective maintenance problem, where the optimization problem is distributed among all the DTs of various components and assembly of the system. These DTs at the lower hierarchy are aware of their own parameters, such as failure and repair distribution parameters, costs, etc., and solve the optimization problem for the localized objective function related to the component or assembly. Such objective functions are part of the core DT domains. It generates their preferences for maintenance activities. Later, the maintenance planning DT has the responsibility to communicate and acquire certain preferences from DTs of all the components and assemblies at the lower hierarchy and work them out for the global objective function for the selective maintenance problem. In this way, the developed methodology splits the larger combinatorial optimization problem into several smaller problems and distributes them to the respective DTs in the network. As all the smaller computation problems are getting solved in parallel, it takes significantly less time. Also, due to the smaller problem size, local-level exploration is better compared to exploration done in complex centralized problems. Furthermore, the maintenance planning DT also needs to solve the problem with already evaluated few feasible solutions only, which enables the optimal solution to be found in significantly less time. In order to prove the hypothesis, the developed distributed selective maintenance optimization algorithm is demonstrated on a benchmark problem of a coal transportation system (Jiang and Liu 2020).

Firstly, we envisioned the coal transportation system to be a CPS, where all the assemblies and the corresponding subassemblies are equipped with their respective DTs and connected in a network. The benchmark system consists of five [Mi ($i = 1, 2 \ldots 5$)] non-identical assemblies (Feeder1, Conveyer1, Stack Reclaimer, Feeder2, Conveyer2) in series with corresponding subassemblies (SA) in parallel Cij where ($j = 1, 2 \ldots Ci$), where Ci is the number of components in assembly Mi. The CPS structure, along with the system configuration of the coal transportation system, is depicted in Fig. 8.2.

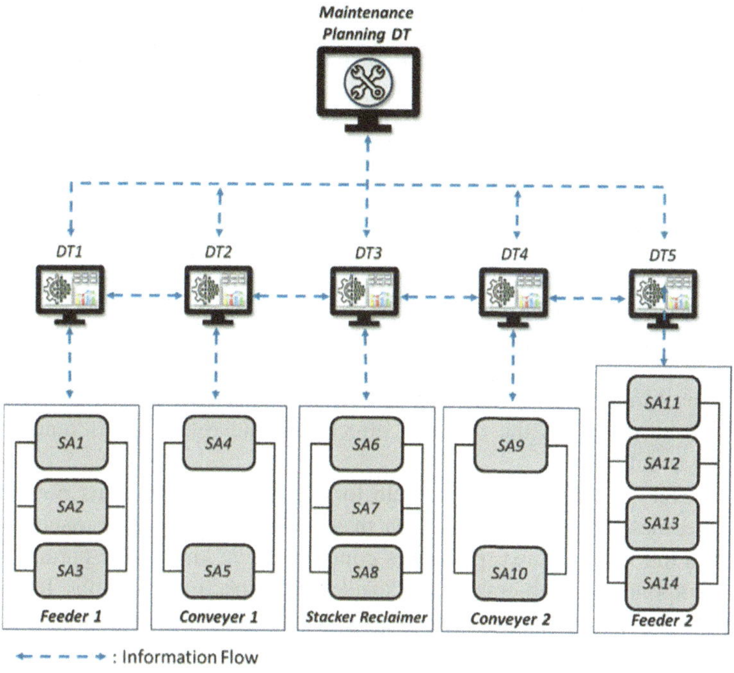

Fig. 8.2 CPS structure of coal transportation system

As discussed above, the overall goal of selective maintenance optimization is to find the best possible maintenance schedule that uses the least maintenance resources while ensuring the minimum required reliability to complete the next mission successfully. Therefore, the objective function of the considered problem is set to minimize maintenance cost and duration while satisfying all of the constraints listed below.

$$Minimize\ TMC = \sum_{i=1}^{M} . \sum_{j=1}^{Ci} AMC_{ij}$$

Subject to:

$R(T) \geq R'(T)$

$TMT \leq Do$

$TMC \leq Co$

$0 \leq Cij \leq (N - 1)$

Where,

T: Mission Duration

R(T): System's Mission Reliability at time T

R'(T): System's Target Mission Reliability at time T

AMC: Actual Maintenance Cost of Asset

TMC: Total Maintenance Cost for the selected SM Schedule

TMT: Total Maintenance Time for the selected SM Schedule

Do: Total intermission duration available

Co: Allocated Mission Budget

N: Number of maintenance actions/ levels (five in this case).

In order to distribute the problem effectively, the roles of different DTs are prescribed as follows:

DT of every assembly is responsible for the identification of its own maintenance schedules through the utilization of selfish algorithms. The assembly level DT, within the local context, possesses comprehensive knowledge pertaining to a certain component or sub-component inside the respective assembly. It conducts assessments on several factors, like maintenance cost, maintenance duration, and post-maintenance reliability. The DT at this local level works towards identifying the most favorable schedules by considering the objective function associated with each of them. The assembly DT passes the highest-priority maintenance schedules to the maintenance planning DT at the subsequent hierarchical level in accordance with the predetermined priority logic.

The maintenance department DT performs the final stage of the optimization. It receives a list of top maintenance schedules from several assembly-level DTs in its own hierarchy. It creates an enumeration of all possible actions and greedily finds schedules that ensure the mission target reliability constraints are met along with other constraints at the system level. From a list of possible favorable schedules, it then returns the one with the least maintenance cost.

The stepwise algorithm to solve the considered problem is:

1. Input and process system metadata like the age of various parts or subcomponents, their reliability parameters, respective maintenance duration, and cost.
2. Identify the system configuration and the hierarchical structure of DT in the CPS network.
3. Decide the number and type of distribution levels and participating DTs. This comes from the learning capability of DT. (However, in the current example, it is pre-fixed.)
4. At the assembly level, DTs apply a selfish algorithm for selective maintenance schedule estimation and pass top X% schedules satisfying the objective function and constraints to the corresponding maintenance planning level DT. Initially, X is randomly decided. Over a period of time, DTs will learn from the various scenarios and optimize X for particular problem cases.

5. At the maintenance planning level DT, apply a greedy algorithm for selective maintenance schedule estimation and prepare a priority list of possible maintenance schedules such that all constraints are satisfied.
6. Return the optimal maintenance schedule for given constraints as per the objective function.

If the system follows a conventional centralized approach, it must enumerate through 5^{14} possible combinations. Which will take huge computation time, making it computationally infeasible. Using evolutionary algorithms like Genetic Algorithm (GA), the issue of computation time could be solved but with a compromise to the optimality, as it could not explore the entire solution space. Instead, by applying the presented DT-based distributed approach, optimality can be achieved within significantly less computation time when compared to centralized approaches using enumeration or evolutionary algorithms. Since all the DTs can compute the problem in parallel, and their problem is converted for a localized objective function, every DT needs to compute for the 5^3, 5^2, 5^3,5^2,and 5^4 maintenance schedules, respectively.

We developed two different approaches in the distributed algorithm, one where enumeration is used at every DT (D_E) and another where GA is used at every DT (D_GA) to solve their respective problems. For a particular case, we applied the presented distributed algorithm on the benchmark problem and found that the presented approach results in optimality in much less computation time when compared with the centralized approach where the GA (C_GA) is employed. For the considered case, as the number of subassemblies is limited to 14, distributed enumeration performed better, but as the number of subassemblies increases, the distributed GA approach is expected to perform better. The results regarding the computation time (ttc) and maintenance cost (mta_cost) in all three cases are depicted in Fig. 8.3.

To assess the consistency of the presented distributed approach, we created 14 different cases by altering the age of the subassemblies and probability distribution parameters. On applying the presented approach to all 14 cases, we found that the distributed enumeration approach consistently performed better. Figure 8.4 shows the comparison of computation time required by each approach in all 14 cases. Also, for all 14 cases, the proposed decentralized algorithms give equal or better performance for the objective function.

It is evident that DT-based distributed approaches produce high-quality results more quickly. As a result, such an approach is ideally suited for dynamic decision-making, which is a crucial necessity for the development of next-generation CPS. The proposed method better aligns with the CPS structure since it takes advantage of distributed data processing and individual systems or components' independent decision-making.

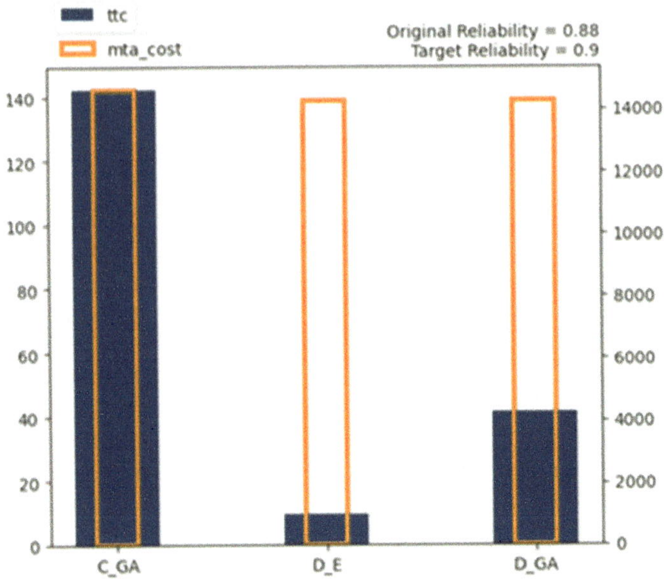

Fig. 8.3 Comparison of computation time and solution quality

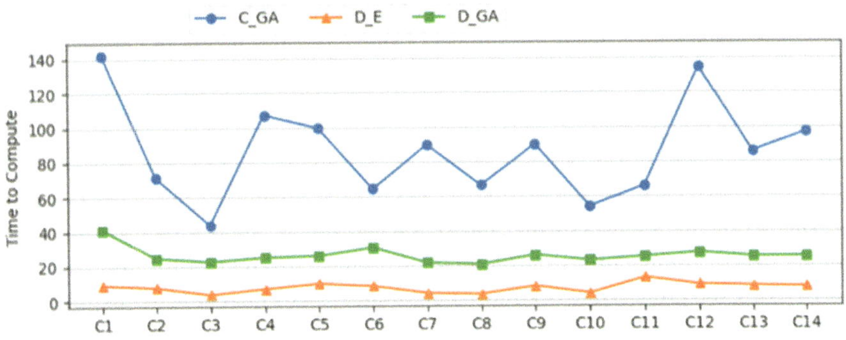

Fig. 8.4 Case-wise comparison of computation time

Conclusion

This research study has highlighted several areas in the RAMS domain where DT-enabled intelligent assets have the potential to significantly impact how operations and maintenance are carried out. This paper dwelled on some of the most important characteristics of DTs, such as real-time synchronization, secure communication, and powerful analytical and modeling capabilities from a RAMS standpoint. These capabilities have the potential to transform traditional reliability engineering and maintenance management techniques across sectors. Furthermore, our research has expanded to a variety of application domains where DTs are set to make significant contributions to RAMS techniques. We have identified key domains where DTs have the potential to optimize asset performance and reliability, ranging from decentralized decision-making for selective maintenance planning to the integration of augmented and virtual reality technologies for maintenance assistance; and from the reliability design process to transfer learning-based prognostics and health management. It can be concluded that DT disrupts reliability testing, maintenance, reliability assessment and provides real-time solutions with ease, extracting greater RAMS utilization. Demonstrating the capabilities of DTs in CPS, we presented a detailed case study on DT based decentralized approach for selective maintenance optimization. This case shows how the capabilities of DTs in CPS can be leveraged to effectively handle complex problems like selective maintenance optimization with decentralized decision making. This aggregated study on multiple developments of the DT towards the betterment of RAMS practices serves as a springboard for future research endeavors in the field of DT-enabled RAMS.

Acknowledgements All the authors would like to acknowledge IITI DRISHTI CPS Foundation, IIT Indore, India, for providing support to the presented research work.

References

EP Editorial Staff (2022). Improve reliability with digital twins. Efficient Plant

Eswaran M, Bahubalendruni MVAR (2022) Challenges and opportunities on AR/VR technologies for manufacturing systems in the context of industry 4.0: a state of the art review. J Manuf Syst 65:260–278. https://doi.org/10.1016/j.jmsy.2022.09.016

Hribernik K, Cabri G, Mandreoli F, Mentzas G (2021) Autonomous, context-aware, adaptive digital twins—State of the art and roadmap. Comput Ind 133:103508. https://doi.org/10.1016/j.compind.2021.103508

ISO Standard (2021) ISO 23247-1:2021—Automation systems and integration—Digital twin framework for manufacturing—Part 1: Overview and general principles

Jay L, Jaskaran S, Moslem A (2023) Industrial artificial intelligence. ArXiv, 2019. Accessed 10 Oct 2023. https://arxiv.org/abs/1908.02150

Jeon B, Yoon JS, Um J, Suh SH (2020) The architecture development of Industry 4.0 compliant smart machine tool system (SMTS). J IntellManuf 31(8):1837–1859. https://doi.org/10.1007/s10845-020-01539-4

Jiang T, Liu Y (2020) Selective maintenance strategy for systems executing multiple consecutive missions with uncertainty. Reliab Eng Syst Saf 193(2006):106632. https://doi.org/10.1016/j.ress.2019.106632

Khalyasmaa AI, Stepanova AI, Eroshenko SA, Matrenin PV (2023) Review of the digital twin technology applications for electrical equipment lifecycle management. Mathematics 11(6). MDPI. https://doi.org/10.3390/math11061315

Mohamed-Hedi K, Brigitte C-M, Noureddine Z (2011) A formal ontology for industrial maintenance. Terminology & Ontology : Theories and applications. In: TOTh conference, Annecy, France, pp 1–20

Montero Jiménez JJ, Vingerhoeds R, Grabot B, Schwartz S (2023) An ontology model for maintenance strategy selection and assessment. J Intell Manuf 34(3):1369–1387. https://doi.org/10.1007/s10845-021-01855-3

Palau AS (2019) Distributed collaborative prognostics. PhD Thesis, University of Cambridge. Distributed Collaborative Prognostics (core.ac.uk). Accessed 5 Oct 2023

Upasani K, Bakshi M, Pandhare V, Lad BK (2017) Distributed maintenance planning in manufacturing industries. Comput Ind Eng 108:1–14. https://doi.org/10.1016/j.cie.2017.03.027

Wu Y, Zhang K, Zhang Y (2021) Digital twin networks: a survey. IEEE Internet Things J 8(18):13789–13804. https://doi.org/10.1109/JIOT.2021.3079510

Chapter 9
Software Technology Management Under Stochastic Environment

P. K. Kapur and Avinash K. Shrivastava

Introduction

Technological advancements have made our life easy and it has reduced our daily tasks by many folds. Software systems are the backbone behind all these advancements. For the smooth functioning of the software systems, understanding and management of software reliability is the key. Software reliability is a critical aspect of software engineering, as the dependability and performance of software systems are essential for various industries and applications. Predicting and managing software reliability growth over time has become a significant challenge. Also, we have seen the massive evolution of software technology in the last few decades (Shrivastava et al. 2015). Over the last five decades, several time-dependent software reliability growth models (SRGMs) have been explored to develop software reliability metrics (Kapur et al. 2011a, b; Pham 1999). Researchers have developed different models based on different sets of assumptions (Goel and Okumoto 1979; Yamada et al. 1983; Kapur and Garg 1992, Huang and Lin 2006). Early software reliability models focused on failure rate estimation based on hardware reliability concepts (Kapur et al. 1999). The need for specific software reliability models became apparent as software systems grew in complexity. Musa's pioneering work on software reliability modelling introduced the idea of a software fault detection process. The initial models were relatively simple, but the concept of considering fault detection in software reliability growth was established (Musa et al. 1987).

Another direction of research in the field of software technology is release time management. Researchers utilized the SRGMs to determine the optimal launch time of the software and its reliability level at the time of release (Kapur et al. 2011a, b).

P. K. Kapur (✉)
Amity Center for Interdisciplinary Research, Amity University, Noida, U.P., India
e-mail: pkkapur1@gmail.com

A. K. Shrivastava
International Management Institute Kolkata, Kolkata, West Bengal, India

P. V. Varde et al. (eds.), *Advances in Risk-Informed Technologies*, Risk, Reliability and Safety Engineering, https://doi.org/10.1007/978-981-99-9122-8_9

Further work has been done to model the complexities in the software technology due to the addition of new functionalities. Various frameworks were proposed to predict the software reliability for multi-release of the software (Kapur et al. 2010a, b; Singh et al. 2014; Kapur et al. 2015; Mishra et al. 2018). During the multi-release modelling it was observed that software release and testing time need segregation (Kapur et al. 2015; Singh et al. 2017; Shrivastava et al. 2023). It was also proposed that it is better to release early and continue to test the software in the operational phase (Shrivastava and Sachdeva 2020). During the debugging of faults, it is possible to have different rates of removal (Shrivastava and Kapur 2021). Technical advancements in the software system have raised security concerns as well. Research related to software vulnerability modelling is one of the niche areas in which frameworks similar to SRGMs are applicable and it has attracted researchers to do further research to find meaningful insight (Kapur et al. 2015; Shrivastava et al. 2021; Sharma et al. 2023).

The software debugging process is influenced by a variety of factors, including testing resources, testing tools, and tester skills (Lo and Huang 2006; Zhu and Pham 2018; Xie et al. 2022). A perfect debugging setup is an ideal scenario. However, errors might occur throughout the fault rectification procedure, resulting in an imperfect paradigm. Many studies have already taken into account the imperfect debugging phenomena while estimating the reliability growth trend of software products (Obha 1989; Lin 2011; Kapur et al. 2011a, b; Kumar et al. 2016; Khurshid et al. 2019; Saraf et al. 2020; Khurshid et al. 2022). Some models assume that once a failure is identified, the faults that caused the failure are instantly repaired and no new errors are introduced i.e. perfect debugging (Pham 2006; Kapur et al. 2011a, b; Kapur et al. 2015; Saraf et al. 2021). Other models take into account imperfect debugging (Yamada and Osaki 1993; Huang 2011), i.e., during the fault removal process new faults are introduced into the software system. Ohba and Chou (1989) were the first to change the growth models' assumption of perfect debugging to account for error generation. They used a real-world project to show the applicability of their concept. Yamada et al. (1992) and Kapur and Garg (1990) also created NHPP-based models that included the idea of imperfect debugging. Pham and Zhang (1997) used the learning phenomena to simulate the defect detection rate when imprecise debugging was present. Furthermore, Pham et al. (1999) created the reliability growth model, which incorporates poor debugging with linearly rising fault content across the testing time. Zhang et al. (2003) modelled the reliability growth pattern using an incomplete debugging framework and incorporated fault elimination efficiency. Kapur et al. (2011a, b) developed a generalized SRGM with a unified approach that incorporates the error generation process. To represent actual scenarios of fault elimination behaviour, they investigated several distribution functions. Aktekin and Caglar (2013) established a probabilistic reliability growth model that incorporates the multiplicative failure rate in another investigation. They used the Bayesian technique to explain the imperfect debugging phenomenon under uncertain environment.

The Musa-Okamura model (also known as the Musa-Okamura-Hagadorn model) introduced the concept of debugging and fault removal rate as an integral part of the reliability growth process. The incorporation of fault detection and correction rates made these models more realistic and useful for practical software development.

However, most of the existing models presume that problems would be rectified instantly when they are recognized. However, this is not practical, and discovered flaws will grow increasingly harder to rectify as testing proceeds. As a result, it is critical to construct software reliability models from the standpoint of the fault correction process, i.e., to prioritize modelling the fault correction process above modelling the problem detection process.

Debugging delay is a significant component that impacts the fault removal process during software testing. A detected fault must be reported, diagnosed, and eventually fixed, and a debugging delay is the amount of time it takes to correct a defect once it has been identified. In the software reliability growth models, debugging delay is not insignificant. Certain scholars have developed certain coupled models of fault detection process (FDP) and fault correction process (FCP) that incorporate the debugging delay (Li and Pham 2021; Zhang et al. 2017; Peng and Liu 2017; Cinque et al. 2017).

Schneidewind (1975) proposed a fault-correction model with a constant time delay in the fault-detection process to first represent the software rectification process alongside the software detection process. The concept of Schneidewind was then expanded by Xie and Zhao (1992) from a constant time delay to a time-dependent delay function. Later, Schneidewind (2002) extended his original model by including a random variable for the time delay that follows an exponential distribution. Lo and Huang (2006) developed a comprehensive framework for modelling software fault detection and rectification procedures and demonstrated that the suggested methodologies may encompass various existing NHPP-based SRGMs. Xie et al. (2007) suggested another distributed correction time model to offer more flexible modelling of corrective processes while Peng et al. (2014) added a testing effort function and imprecise debugging into the time delay function. Yang et al. (2016) suggested a hybrid genetic algorithm-based approach that uses model mining techniques to uncover failure correlations and optimal model parameters. Wanget al. (2016), Liu et al. (2016) investigated a framework of software reliability models that included both detection and correction processes together. Though these efforts have brought software dependability models closer to reality, they still presume that all software errors are of the same sort. That is, all problems are detected and corrected at the same pace. This assumption, however, may not be feasible. Due to human variables and the unpredictability of the field environment, the software contains various defects and detecting/correcting different problems may necessitate different efforts. As a result, errors may have variable detection and rectification rates. If this assumption is not taken into consideration, the findings of SRGM may be deceptive.

During the debugging process, Wang and Wu (2016) included a non-linear growth in the introduced flaws. Chatterjee and Shukla (2016a, b) provided an extended framework for the software reliability growth model (SRGM). In modelling the problem detection and correction process, they used the concepts of imprecise debugging, change-point, and problem Reduction Factor (FRF). Under the uncertainties of the field environment and an unsatisfactory debugging procedure, Li and Pham (2017 2019) established the testing-coverage dependent software reliability model. They used a randomly distributed variable to investigate uncertainty in operational situations. Rani and Mahapatra (2019) suggested a Neural Network (NN) technique for analyzing software reliability when errors are generated. They used the neural network (NN) technique to more precisely evaluate and forecast software dependability. By merging imperfect debugging and coverage variables, Chatterjee and Shukla (2019) developed a unified way to modelling software reliability growth. Huang et al. (2022) suggested a reliability model that takes into account the learning effect, different types of mistakes, change points, and incomplete debugging in a recent attempt to expand the viability of SRGMs. All of these current models have demonstrated the importance of imperfect debugging and error creation in predicting dependability accurately. (Tanaka et al. 1992; Shyur et al. 2003; Tamura and Yamada 2021; Yang et al. 2022; Kapur and Shrivastava 2015; Kapur et al. 2017; Shrivastava and Sharma 2022). Nonetheless, all previous research has represented software dependability as a monotonic or distributed function of time throughout an inadequate debugging process. Their concept does not incorporate uncertainty or unpredictability into the error-generating mechanism. These studies assume that the rate of fault introduction remains constant throughout the debugging procedure. As a result, they fail to include dynamicity in the fault introduction process (Bibyan et al. 2023).

Related Work

Early software reliability growth models primarily focused on capturing the increase in reliability by incorporating the number of failures. These models, however, didn't consider the inherent variability in software development and usage (Yamada et al. 2003; Lee et al. 2004). The introduction of stochastic processes and differential equations aimed to overcome this limitation (Tamura and Yamada 2006, Kuo et al. 2006; Koshgoftar et al. 2007). Stochastic Differential Equations (SDEs) provide a framework to model continuous-time processes affected by randomness. SDE-based SRGMs capture the dynamics of software reliability growth by incorporating randomness in the development process, including fault detection and removal rates (Shyur 2003; Tamura et al. 2006; Kapur et al. 2009a, b). Stochastic Differential Equation (SDE)-based software reliability growth models (SRGMs) have emerged as a promising approach to address this challenge. SDE-based SRGMs have found applications in various industries, including aerospace, telecommunications, and finance, where software reliability is critical (Kapur et al.2010a, b). Case studies involving real-world software projects have demonstrated the effectiveness of SDE-based models in predicting reliability growth and estimating the optimal release time for software systems. These models consider factors such as fault detection rates, fault removal efficiency, and system usage patterns (Tamura and Yamada 2010, Khatri et al. 2011; Kuo et al. 2012; Kapur et al. 2012).

In the past, authors have incorporated uncertainty and dynamicity in the fault detection process (Kapur et al. 2009a, b; Tamura and Yamada 2014, 2017, 2019, Singh et al. 2015, 2021; Guo et al. 2019; Zhu and Pham 2020; Chatterjee et al. 2020). Only a few attempts have been made to take into account irregularity in the rate of defect introduction. Wang and Wu (2016) developed the first imperfect software debugging model that includes a fault introduction rate as a variable that varies irregularly over time and fault introduction as a nonlinear technique. The study by Huang et al. (2022) is based on the NHPP and takes into account the phenomena of incomplete debugging, different types of faults, and change points during the testing period in order to increase the practicability of SRGMs. They represented the error fluctuation rate as a sine cyclical function with diminishing fluctuation breadth during the detection period, implying that the impact of rising new mistakes during the testing period will eventually become unnoticeable as the testing time passes. Furthermore, the time required to remove simple or complicated mistakes is assumed to follow distinct truncated exponential distributions. Liu et al. (2022) consider software reliability from a new perspective within the framework of uncertainty theory, which is a new mathematical system distinct from probability theory, and propose for the first time a software belief reliability growth model (SBRGM) based on uncertain differential equations. Based on this SBRGM, the features of critical software reliability metrics are examined using the belief reliability theory, a novel reliability theory. Based on the belief reliability theory, Liu and Kang (2022) developed an imperfect debugging software belief reliability growth model using the uncertain differential equation within the framework of uncertainty theory and investigates properties of essential software belief reliability metrics, namely belief reliability, belief reliable time, and mean time between failures. In this model, estimates for unknown parameters are obtained. All of the models described assumed that the fault introduction rate is either a power function or an exponential function of time. Singhal et al. (2023) suggested a reliability growth model for OSS by introducing dynamicity into the debugging process in response to this gap. They developed stochastic differential equation-based analytical models to represent the instantaneous rate of error generation by taking fault introduction rate as exponential and Erlang distribution functions. They incorporated an S-shaped distributed bug introduction rate and dynamicity in the bug generation phenomenon. In this work, we have proposed a generalized framework for fault introduction under a stochastic environment by considering it as a two-stage model. It can be easily verified that all the existing models are some special cases of our proposed model.

The rest of this paper is structured as follows. In Sect. 9.3, we first provide a brief overview of the existing fault-detection models' assumptions and fault-correction process, then construct a relationship between the number of detected and corrected faults, and finally present the proposed model, which includes a two-stage stochastic fault introduction rate. Section 9.4 compares the fitting and prediction performance of our model to other current SRGMs on two sets of software failure data. Finally, Sect. 9.5 summarizes the suggested study and suggests future research directions.

Model Development

In this section we have firstly described the notations and assumptions used to develop the proposed model.

Notations

$m(t)$	Expected number of faults removed by time 't'
β	Learning parameter
α	Error generation rate
a	Initial fault content
b	Fault removal rate
σ	Fluctuation rate in SDE
a_1	Total number of faults introduced due to fluctuation
$\gamma(t)$	Standard Gaussian white noise
$a(t)$	Time-dependent stochastic fault content function

Assumptions

The assumptions on which the proposed methodology is based are as follows:

1. The fault removal process follows the Non-homogeneous Poisson Process (NHPP).
2. The failure occurrence in the system is independent and identically distributed over the period with a cumulative distribution function as F(t) = P(X $\leq$ t) = f(x)dx, where P(·) represents the probability of failure occurrence.
3. The number of system failures is determined by the number of unidentified software faults at testing time t.
4. The fault removal phenomena is presumed to be imperfect i.e. some new faults may be introduced into the system during the debugging process of eliminating the inherent faults.
5. When faults are discovered, the testing team corrects them right away. New flaws, however, might be introduced at that time.
6. The rate of fault introduction during the debugging procedure varies randomly over time. Over the testing time, the fault content rises non-linearly and in a non-deterministic manner.

Kapur et al. (2011a, b) categorized faults into two types namely dependent and independent faults. Faults which are recognized on a failure are called as independent faults while the faults removed in addition are named as dependent faults. They proposed the following equation to distinguish between these two types of faults.

$$\frac{f(t)}{1-F(t)} = p + qF(t) \tag{9.1}$$

where $f(t)$ is the probability density function of detection; $F(t)$ is a cumulative distribution function i.e., $F(t) = \frac{m(t)}{a}$; p symbolizes the fault detection rate for independent faults and q denotes the fault detection rate for dependent faults which is considered constants for this formulation. Additionally, let a denote the fault content. Then, Kapur and Garg Model (1992) (fault removal phenomenon) represents the cumulative number of faults which is attained by the following differential equation (DE):

$$\frac{dm(t)}{dt} = \left(p + q\frac{m(t)}{a}\right)(a - m(t)) \tag{9.2}$$

where, m(t) defines the cumulative number of faults by time t, and (a − m(t)) denotes the residual number of faults.With the initial condition m(0) = 0. The above DE (Eq. 9.2) is further solved, to accomplish the mean value function i.e.;

$$m(t) = a\left(\frac{1 - e^{-(p+q)t}}{1 + \left(\frac{q}{p}\right)e^{-(p+q)t}}\right) \tag{9.3}$$

Equation (9.3) shows an S-Shaped form over the entire software lifespan. According to Kapur et al. (2011a, b) the alternative differential equation can be expressed as follow:

$$\frac{dm(t)}{dt} = b(t)(a(t) - m(t)) \tag{9.4}$$

here $b(t)$ signifies the time-dependent detection rate and can be expressed as:

$$b(t) = \frac{b}{1 + \beta e^{-bt}} \tag{9.5}$$

In Eq. (9.5), b is the detection parameter and β denotes the learning parameter. In accumulation, Kapur et al. (2004) presumed that fault content of a software develops exponentially.

$$\textbf{Case I: } a(t) = ae^{\alpha t}, \ \alpha > 0; \ \text{Kapur et al. (2006)} \tag{9.6}$$

where a is the original fault content and α denotes the error generation rate. Thus, with the seed value $m(0) = 0$, Eq. (9.4) becomes:

$$m(t) = a\left(\frac{b}{\alpha + b}\right)\left[\frac{e^{\alpha t} - e^{-bt}}{1 + \beta e^{-bt}}\right] \tag{9.7}$$

where $F(t) = \left(\frac{e^{\alpha t} - e^{-bt}}{1+\beta e^{-bt}}\right)$, Considering, (i) $b = p + q$, (ii) $\beta = \frac{q}{p}$, and (iii) $\alpha = 0$, i.e. fault content is constant. We can conclude that the K-G model alternatively comes from the Kapur et al. (2004) model.

Authors have considered other cases also, which are as follows:

Case II: $a(t) = a(1 + \alpha t)\alpha > 0$

$$b(t) = \frac{b}{1 + \beta e^{-bt}}$$

$$m(t) = \left(\frac{a}{1 + \beta e^{-bt}}\right)\left[\left(1 - e^{-bt}\right)\left(1 - \frac{\alpha}{\beta}\right) + \alpha t\right]$$

Case III: $a(t) = a + \alpha m(t)\alpha > 0$

$$b(t) = \frac{b}{1 + \beta e^{-bt}}$$

$$m(t) = \left(\frac{a}{1 - \alpha}\right)\left[1 - \left(\frac{(1 + \beta)e^{-b(1-\alpha)t}}{1 + \beta e^{-bt}}\right)\right]$$

Case IV: $a(t) = a(1 + \alpha t)\alpha > 0$

$$b(t) = \frac{b^2 t}{1 + bt}$$

$$m(t) = a\left(1 + \alpha t - \frac{1 + bt}{e^{bt}}\right) - \frac{a\alpha(1 + bt)}{be^{bt}}\left(\ln(1 + bt) + \sum_{i=0}^{\infty} \frac{(1 + bt)^{(i+1)} - 1}{(i + 1)!(i + 1)}\right)$$

Case V: $a(t) = a + \alpha m(t)\alpha > 0$

$$b(t) = \frac{b^2 t}{1 + bt}$$

$$m(t) = \left(\frac{a}{1 - \alpha}\right)\left[1 - (1 + bt)^{1-\alpha} e^{-b(1-\alpha)t}\right]$$

Proposed Model

The fault removal rate equations presented in Eqs. (9.2) and (9.4) have a certainty interpretation. As a result, their behaviour may be predicted in a deterministic manner.

Testing proceeds in an unpredictable manner due to the non-deterministic behaviour of different aspects such as testing effort expenditure, testing efficiency and skill, testing technique and strategy. As a result, the fault content of software is characterized in probabilistic terms, and the differential equation may be constructed as follows based on the above-mentioned assumption (Eq. 9.2):

$$\frac{dm(t)}{dt} = b(t)[Ea(t) - m(t)] \tag{9.8}$$

Here $m(t)$ denotes the cumulative number of faults in the software by the time t, $b(t)$ denotes the hazard rate function for fault removal; $a(t)$ represents the time dependent stochastic fault content. So, hazard rate function is described as follow:

$$b(t) = \frac{f(t)}{1 - F(t)} \tag{9.9}$$

Again, the detection rate $b(t)$ following logistic rate can be advocated

$$b(t) = \frac{b}{1 + \beta e^{-bt}} \tag{9.10}$$

as pre-arranged by Kapur et al. (2004). Here b indicates the detection or measure parameter and β signifies the learning parameter. Increase value of b signifies the decrease the value of β and hence make the possibility of rapid fault removal phenomenon. Thus, (Eq. 9.10 in Eq. 9.8), the instantaneous fault removal rate equation converts to:

$$\frac{dm(t)}{dt} = \frac{b}{1 + \beta e^{-bt}}(E(a(t) - m(t)) \tag{9.11}$$

Equation (9.11) portrays the proposed fault removal rate equation with variable fault content. As per sixth postulation, the DE representing total fault content can be stated using Eq. (9.12):

$$\frac{da(t)}{dt} = \alpha(t)[(a + a1) - a(t)] \tag{9.12}$$

where $\alpha(t)$ denotes time dependent function in respect of fault content. $a1$ is the upper bound of increase in the fault content. In this proposed model, the behaviour of fault removal is framed using Brownian motion process during the software development life cycle. And fault content $a(t)$ is a random variable following the stochastic process. Further, during the fault removal process the introduction of fault can either be slow or fast depending on the testing environment, skills of the testing team, or the testing resources. To incorporate the different rate of fault introduction during the fault removal, let us assume that, $h(t)$ and g(t) are the probability density functions and

$H(t)$ *and* $G(t)$ are the corresponding cumulative distribution functions of low and high rate of fault introduction, then the mathematical expression for fault content using the postulate (9.4) is:

$$\alpha(t) = \frac{h(t) * g(t)}{1 - H(t)\Theta G(t)} + \sigma\gamma(t) \tag{9.13}$$

$\frac{h(t)*g(t)}{1-H(t)\Theta G(t)}$ time dependent rate of fault content, σ constant representing the scale of irregular fluctuation, and $\gamma(t)$ standard gaussian white noise that is stochastic in nature. On substituting Eq. (9.13) in Eq. (9.12):

$$\frac{da(t)}{dt} = \left(\frac{h(t) * g(t)}{1 - H(t)\Theta G(t)} + \sigma\gamma(t)\right)(a + a1 - a(t)) \tag{9.14}$$

Equation (9.14) symbolizes the SDE, a modified of ordinary differential equations that are parameterized by Weiner processes. Itô stochastic calculus is applied to formulate the equation.

$$da(t) = \left(\frac{h(t)*g(t)}{1-H(t)\Theta G(t)} - \frac{\sigma^2}{2}\right)(a + a1 - a(t))dt + \sigma(a + a1 - a(t))dW(t)\Big\} \tag{9.15}$$

In this Eq. (9.15), $W(t)$ symbolizes a Brownian motion or wiener process and it hold the following axioms:

(i) Continuous process with $W(0) = 0$
(ii) Have independent increments $\forall t > 0$
(iii) Have Gaussian increments, i.e., $W(t+dt) - W(t) \sim N(0, dt)$; where $N(0, dt)$ is a normal distribution centred at zero.

Now, if x is a random variable then, $f_{w_t}(x) = \frac{1}{\sqrt{2\pi t}} e^{-\frac{x^2}{2t}}$.

Using the seed value, $a(0) = 0$ Eq. (9.15) can be combined using the Itô formula to obtain the cumulative fault content function:

$$a(t) = a + a1(1 - (1 - H(t)\Theta G(t))e^{-\sigma W(t)} \tag{9.16}$$

From Eq. (9.16), it can be derived that $a(0) = a$ when $t = 0$ and $a(\infty) = a + a1$, when $t \to \infty$. That is, the fault content at entry time is a and eventual fault content over its life cycle will be $a + a1$. Considering expectation on both sides, Eq. (9.16) is assumed as:

$$E[a(t)] = a + a1\left((1 - (1 - H(t)\Theta G(t))e^{\frac{\sigma^2 t}{2}}\right) \tag{9.17}$$

Equation (9.17) belong the uncertainty and random fluctuations in fault introduction rate. It was designed using two types of distribution functions to represent two different behaviour of fault introduction function in the system.

Case I: In this case, $H(t) \sim exp(\alpha) and G(t) \sim 1(t)$

It defines a continuous process where increments occur independently at a constant rate.

$$H(t) = \left(1 - e^{-\alpha t}\right) \text{ and } G(t) = 1 \tag{9.18}$$

Here, α is a scalar parameter signifying the introduction rate of fault content. Thus, using Eq. (9.18), the expected value of fault content is:

$$E[a(t)] = a + a1\left((1 - e^{-\alpha t + \frac{\sigma^2 t}{2}}\right) \tag{9.19}$$

Case II: In this Case, $H(t) \sim exp(\alpha)\ and\ G(t) \sim exp(\alpha)$

$$H(t) = \left(1 - e^{-\alpha t}\right) \text{ and } G(t) = (1 - e^{-\alpha t}) \tag{9.20}$$

Thus, the expected value becomes:

$$E[a(t)] = a + a1(1 - (1 + \alpha t)e^{-\alpha t + \frac{\sigma^2 t}{2}}) \tag{9.21}$$

Equation (9.21) symbolizes the S-shaped pattern of fault content function.Using Eqs. (9.19) and (9.21), the differential Eq. (9.11) representing fault removal rate function under $t = 0$, $N(t) = 0$ to attain the mean value function. In Table 9.1, two diverse stochastic software reliability growth models are reported using the proposed modelling framework.

Here it is important to note that if we consider H(t) as identity function then the proposed model will result in the model proposed by Singhal et al. (2023). i.e. From Eq. (9.17) we have

$E[a(t)] = a{+}a1\left((1 - (1 - H(t)\Theta G(t))e^{\frac{\sigma^2 t}{2}}\right)$ By taking H(t) = I(t) it will become

Table 9.1 Proposed stochastic models

SDE based models	Fault content function	Mean failure function
Model 1	$E[a(t)] = a + a1\left((1 - e^{-\alpha t + \frac{\sigma^2 t}{2}}\right)$	$m(t) = \frac{1}{1+\beta e^{-bt}}\left((a + a1)\left(1 - e^{-bt}\right) + \frac{2ba1}{-\sigma^2+2\alpha-2b}\left(e^{-\alpha t + \frac{\sigma^2 t}{2}} - e^{-bt}\right)\right)$
Model 2	$E[a(t)] = a + a1(1 - (1 + \alpha t), e^{-\alpha t + \frac{\sigma^2 t}{2}})$	$m(t) = \frac{1}{1+\beta e^{-bt}}\left((a + a1)\left(1 - e^{-bt}\right) + \frac{2ba1}{-\sigma^2+2\alpha-2b}\left(\left(1 + \alpha t - \frac{2\alpha}{\sigma^2-2\alpha+2b}\right)e^{-\alpha t + \frac{\sigma^2 t}{2}} - \left(1 - \frac{2\alpha}{\sigma^2-2\alpha+2b}\right)e^{-bt}\right)\right)$

$E[a(t)] = a + a1\left((1-(1-G(t))e^{\frac{\sigma^2 t}{2}}\right)$ which is same as the model proposed by Singhal et al.(2023). Similarly we can derive all the other existing models by taking G(t) as different distribution functions. Therefore, we can say that all the existing models are the particular cases of our proposed model.

Numerical Illustration

Description of Data Sets

For the current investigation, data from the Eclipse open-source project was obtained from publicly available sources. Eclipse is a computer programming integrated development environment (IDE). The dataset is huge and contains a list of all the problems encountered during the software development cycle (Yaghoobi 2020). Overall, 6495 bugs are reported in 147 months from October 2001 to December 2013.

Comparison Criteria

1. The Coefficient of Determination (R^2). It measures the percentage of variations that can be explained by the model, with a higher value indicating a better fit. Note that R^2 is a proportion, thus its value always is between 0 and 1. R^2 is typically defined by (Kapur et al. 2011a, b)

$$R^2 = 1 - \frac{\left(m_i - \widehat{m_i}\right)^2}{\left(m_i - \overline{m}\right)^2}$$

where m_i is the actual value at the time i, $\widehat{m_i}$ is the predicted value at the time i, and $\overline{m}$ is the mean.

2. Root Mean Square Prediction Error (RMSPE): It is a Measure of the Closeness with Which the Model Predicts the Observation. It is Defined by (Kapur et al. 2011a, b)

$$RMSPE = \sqrt{Bias^2 + Variation^2}$$

where $Variation = \frac{1}{(n-1)} \sum_{i=1}^{n} \left[\left(m_i - \widehat{m_i}\right) - Bias\right]^2$

3. Akaike Information Criterion (**AIC**) is defined as (Kapur et al. 2011a, b)

$$AIC = -2LogL + 2 \times N$$

where N is the number of parameters used in the model and L is the value of the likelihood function at its maximum. A lower value represents a more confident predictive validity. AIC considers the degree of freedom by assigning a larger penalty to models with more parameters.

The estimated parameter findings, as well as the comparison criteria, are shown in Table 9.2. The findings show that the proposed models have the lowest RMSE and AIC values while having the greatest R-square values. As a consequence, the parameter estimate results demonstrate that the proposed SRGMs can better match Open Source Software Projects. As a result, imperfect debugging with stochastic consideration is consistent with real-world data from Open Source Software initiatives.

Another notable use of the above-mentioned approach is in technology management. Technology management is the unacknowledged competitive advantage that bridges "the knowledge and practice gap" between science, engineering, and business management (Khalil 2000). Technology Management is a profession that brings together "engineering, science, and management disciplines to plan, develop, implement technological capabilities to shape and accomplish the strategic and operational

Table 9.2 Parameter estimation results and comparison with existing SRGMs

Model	Parameter estimate	Statistical measures		
		RMSE	AIC	R^2
Pham and Zhang (1997)	$a = 6695 c = 44.88 b = 0.035 \beta = 7.471 \delta = 0.999$	117.35	1412.95	0.9969
PNZ (1999)	a = 6286, b = 0.037, β = 8.40, δ = 0.0008	161.23	1504.35	0.9941
Kapur et al. (2006)	a = 6406.026, b = 0.032, β = 5.979, δ = 0.00076	108.93	1389.06	0.9973
Pachauri et al.(2015)	a = 8392, b = 0.125, β = 4.79, δ = 0.158, r = 0.081	108.42	1389.69	0.9973
Chatterjee and Shukla (2019)	a = 10,449, b = 0.026, β = 0.481, K = 0.6121908	113.39	1400.88	0.9971
Yamada and Tamura (2016)	a = 7605 b = 0.024, δ = 0.0037, Ω = 0.977	111.59	1396.16	0.9972
Proposed model 1	$a_0 = 6794 \nu = 0.489 b = 0.026 \beta = 4.121 \delta = 0.204, A = 533$	78.73	1297.62	0.9986
Proposed model 2	$a_0 = 7572 \nu = 0.471 b = 0.041 \beta = 8.56 \delta = 0.13 A = 890$	83.64	1315.40	0.9984

objectives of an organisation." Considering the importance of technology management researchers have developed quantitative models similar to the model developed in the field of software reliability. These models are developed to predict the adopters of technological products. One of the pioneer work was proposed by Frank M Bass who proposed Bass model (1969) and termed it as innovation diffusion model. This well-known model correctly showed the gradual growth in the number of adopters induced by both mass communication and interactions between users and potential users. Few other diffusion models were proposed by (Kapur et al. 2004) which underlie the Bass model from both external and internal point of view and have been efficient in modelling customer adoption process but still the Bass model supersedes them. We are also working on to develop similar quantitative model similar to what we have proposed in Sect. 9.3 to capture the dynamic market condition under the assumption that the adoption of the product follows a two stage i.e. awareness and then purchase. Our motive is to develop a generalized innovation diffusion model considering dynamic market under stochastic environment.

Conclusions

Evolution of software technology has been instrumental in today's development. Need of technically advanced software is never going to stop and hence its becomes important to ensure the reliability of the software system. Better management of software technology can provide many useful insights including reliability and launch time of software. Stochastic Differential Equation-based Software Reliability Growth Models have emerged as powerful tools for predicting and managing software reliability growth. By incorporating randomness and continuous-time dynamics, these models offer a more realistic representation of software development processes. Despite challenges in parameter estimation and adapting to dynamic development environments, SDE-based SRGMs continue to advance our understanding of software reliability and contribute to better software engineering practices. Recent advancements in SDE-based SRGMs include incorporating more realistic assumptions about the underlying stochastic processes, considering multiple fault types, and adapting models to agile software development practices. In this work we have proposed a generalized framework for SDE based SRGM by considering fault introduction rate as a two-stage model and compared it with several existing models. Our model provides better results in comparison to the other existing models. It can also be seen that all the existing SDE based imperfect debugging SRGMs are the special case of our proposed model.

We have also discussed the implementation of similar modelling frameworks in the field of software technology viz. software security and diffusion of technology. Importance of modelling frameworks in software technology to gain meaningful insights like software launch time, testing stop time and multi up-gradation is also briefly discussed. Future research may explore the effectiveness of early defect detection techniques in improving overall software reliability. Also, machine learning and data-driven approaches started influencing software reliability modeling, providing opportunities for predicting defects and guiding correction efforts and the focus has shifted towards assessing the effectiveness of different detection correction strategies in reducing defects and enhancing software quality. Future research can aim to address the challenges related to the heterogeneity of software systems and the impact of various factors (such as software complexity, development methodologies, and team size) on reliability growth. Advances in artificial intelligence and machine learning have the potential to revolutionize the way software reliability models are developed and applied. Similar new research directions can be explored with respect to the other related fields of software technology including information technology and software security.

References

Bibyan R, Anand S, Aggarwal AG, Kaur G (2023) Multi-release software model based on testing coverage incorporating random effect (SDE). MethodsX 10:102076

Chatterjee S, Shukla A (2016a) Modellingand analysis of software fault detection and correction process through Weibull-type fault reduction factor, change point and imperfect debugging. Arab J Sci Eng 41:5009–5025

Chatterjee S, Shukla A (2016b) Modelling and analysis of software fault detection and correction process through Weibull-type fault reduction factor, change point and imperfect debugging. Arab J Sci Eng 41(12):5009–5025

Chatterjee S, Shukla A (2019) A unified approach of testing coverage-based software reliability growth modelling with fault detection probability, imperfect debugging, and change point. J Softw Evol Process 31(3):e2150

Chatterjee S, Chaudhuri B, Bhar C, Shukla A (2020) Optimal release time determination using FMOCCP involving randomized cost budget for FSDE-based software reliability growth model. Int J Reliab Qual Saf Eng 27(01):2050004

Cinque M, Cotroneo D, Pecchia A, Pietrantuono R, Russo S (2017) Debugging-workflow aware software reliability growth analysis. Softw Test Verif Reliab 27(7):e1638

Guo C, Liu M, Xie M (2019) A stochastic differential equation model for software reliability growth considering imperfect debugging. Reliab Eng Syst Saf 191:106575

Huang CY, Lin CT (2006) Software reliability analysis by considering fault dependency and debugging time lag. IEEE Trans Reliab 55(3):436–450

Huang CY, Lyu MR (2011) Estimation and analysis of some generalized multiple change-point software reliability models. IEEE Trans Reliab 60(2):498–514

Huang YS, Chiu KC, Chen WM (2022) A software reliability growth model for imperfect debugging. J Syst Softw 188:111267

Kapur PK, Goswami DN, Gupta A (2004) A software reliability growth model with testing effort dependent learning function for distributed systems. Int J Reliab Qual Saf Eng 11(04):365–377

Kapur PK, Shatnawi O, Aggarwal A, Kumar R (2009b) Unified framework for developing testing effort dependent software reliability growth models. WSEAS Trans Syst 8(4):521–531

Kapur PK, Basirzadeh M, Inoue S, Yamada S (2010a) Stochastic differential equation-based SRGM for errors of different severity with testing-effort. Int J Reliab Qual Saf Eng 17(03):179–197

Kapur PK, Pham H, Anand S, Yadav K (2011a) A unified approach for developing software reliability growth models in the presence of imperfect debugging and error generation. IEEE Trans Reliab 60(1):331–340

Kapur PK, Pham H, Gupta A, Jha PC (2011b) Software reliability assessment with OR applications, vol 364. Springer, London

Kapur PK, Anand S, Yadav K, Singh J (2012) A unified scheme for developing software reliability growth models using stochastic differential equations. Int J Oper Res 15(1):48–63

Kapur PK, Shrivastava AK (2015) Release and testing stop time of a software: a new insight. In: 2015 4th international conference on reliability, infocom technologies and optimization (ICRITO) (Trends and future directions). IEEE, pp 1–7

Kapur PK, Kumar S, Garg RB (1999) Contributions to hardware and software reliability, vol 3. World Scientific

Kapur PK, Kumar D, Gupta A, Jha PC (2006) On how to model software reliability growth in thepresence of imperfect debugging and error generation. In: Proceedings of 2nd international conferenceon reliability and safety engineering, pp 515–523

Kapur PK, Anand S, Yamada S, Yadavalli VS (2009) Stochastic differential equation-based flexible software reliability growth model. Mathematical Problems in Engineering

Kapur PK, Tandon A, Kaur G (2010) Multi up-gradation software reliability model. In: 2010 2nd international conference on reliability, safety and hazard-risk-based technologies and physics-of-failure methods (ICRESH). IEEE, pp 468–474

Kapur PK, Singh O, Shrivastava AK, Singh JN (2015) A software up-gradation model with testing effort and two types of imperfect debugging. In: 2015 international conference on futuristic trends on computational analysis and knowledge management (ABLAZE). IEEE, pp 613–618

Kapur PK, Panwar S, Shrivastava AK, Khatri SK (2017) Multi-generation diffusion of technology. In 2017 6th International conference on reliability, infocomtechnologiaes and optimization (trends and future directions) (ICRITO). IEEE, pp 31–37

Khalil T (2000) Management of technology: the key to prosperity and wealth creation. McGraw Hill, New York

Khatri SK, Trivedi P, Kant S, Dembla N (2011) Using artificial neural-networks in stochastic differential equations based software reliability growth modeling. J Softw Eng Appl 4(10):596–601

Khoshgoftaar TM, Allen EB, Goel N (2007) A stochastic differential equation model for software reliability growth analysis. J Comput Sci Technol 7(1):1–8

Khurshid S, Shrivastava AK, Iqbal J (2019) Generalized multi-release framework for fault prediction in open source software. Int J Softw Innov (IJSI) 7(4):86–107

Khurshid S, Shrivastava AK, Iqbal J (2022) Generalised multi release framework for fault determination with fault reduction factor. Int J Inf Comput Secur 17(1–2):164–178

Kumar V, Sahni R, Shrivastava AK (2016) Two-dimensional multi-release software modelling with testing effort, time and two types of imperfect debugging. Int J Reliab Saf 10(4):368–388

Kuo SY, Kao TW (2012) A stochastic differential equation model for software reliability growth with testing-effort and learning effects. J Syst Softw 85(6):1280–1290

Lee CH, Kim YT, Park DH (2004) S-shaped software reliability growth models derived from stochastic differential equations. IIE Trans 36(12):1193–1199

Li Q, Pham H (2021) Modelling software fault-detection and fault-correction processes by considering the dependencies between fault amounts. Appl Sci 11(15):6998

Lin CT (2011) Analyzing the effect of imperfect debugging on software fault detection and correction processes via a simulation framework. Math Comput Model 54(11–12):3046–3064

Liu Z, Kang R (2022) Imperfect debugging software belief reliability growth model based on uncertain differential equation. IEEE Trans Reliab 71(2):735–746

Liu Y, Li D, Wang L, Hu Q (2016) A general modelling and analysis framework for software fault detection and correction process. Softw Test Verif Reliab 26(5):351–365

Lo JH, Huang CY (2006) An integration of fault detection and correction processes in software reliability analysis. J Syst Softw 79:1312–1323

Mishra P, Shrivastava AK, Kapur PK, Khatri SK (2018) Modelling fault detection phenomenon in multiple sprints for Agile software environment. Quality, IT and business operations: modelling and optimization, pp 251–263

Musa JD, Iannino A, Okumoto K (1987) Software reliability: measurement, prediction, application. McGraw-Hill, Inc
Obha M, Chou X (1989) Does imperfect debugging affect software reliability growth models. In: Proceedings of the 4th international conference on software engineering, Pittsburg, pp 237–244
Pachauri B, Dhar J, Kumar A (2015) Incorporating inflection S-shaped fault reduction factor to enhancesoftware reliability growth. Appl Math Model 39(5–6):1463–1469
Peng R, Li YF, Zhang WJ, Hu QP (2014) Testing effort dependent software reliability model for imperfect debugging process considering both detection and correction. Reliab Eng Syst Saf 126:37–43
Peng R, Liu J (2017) Simulated software testing process considering debuggers with different detection and correction capabilities. Int J Perform Eng 13(3)
Pham H, Zhang X (1997) An NHPP software reliability model and its comparison. Int J Reliab Qual Saf Eng 4(03):269–282
Pham H, Nordmann L, Zhang Z (1999) A general imperfect-software-debugging model with S-shapedfault-detection rate. IEEE Trans Reliab 48(2):169–175
Pham H (2006) Software reliability modeling. Syst Softw Reliab 153–177
Rani P, Mahapatra GS (2019) A novel approach of NPSO on dynamic weighted NHPP model for software reliability analysis with additional fault introduction parameter. Heliyon 5(7)
Saraf I, Shrivastava AK, Iqbal J (2020) Generalised fault detection and correction modelling framework for multi-release of software. Int J Ind Syst Eng 34(4):464–493
Saraf I, Shrivastava AK, Iqbal J (2021) Effort-based fault detection and correction modelling for multi release of software. Int J Inf Comput Secur 14(3–4):354–379
Schneidewind NF (2002) An integrated failure detection and fault correction model. In: International conference on software maintenance, 2002. Proceedings. IEEE, pp 238–241
Sharma R, Shrivastava AK, Pham H (2023) Software security evaluation using multilevel vulnerability discovery modelling. Qual Eng 35(2):341–352
Shrivastava AK, Kapur PK (2021) Change-points-based software scheduling. Qual Reliab Eng Int 37(8):3282–3296
Shrivastava AK, Sachdeva N (2020) Generalized software release and testing stop time policy. Int J Qual Reliab Manage 37(6/7):1087–1111
Shrivastava AK, Sharma R (2022) Determining optimal release and testing stop time of a software using discrete approach. Int J Softw Innov (IJSI) 10(1):1–13
Shrivastava AK, Ahluwalia AS, Kapur PK (2021) On interdisciplinarity between product adoption and vulnerability discovery modelling. Int J Syst Assur Eng Manage 12:176–187
Shrivastava AK, Sharma R, Pham H (2023) Software reliability and cost models with warranty and life cycle. Proc Inst Mech Eng Part O J Risk Reliab 237(1):166–179
Shrivastava AK, Sharma R, Kapur PK (2015) Vulnerability discovery model for a software system using stochastic differential equation. In: 2015 international conference on futuristic trends on computational analysis and knowledge management (ABLAZE). IEEE, pp 199–205
Shyur HJ (2003) A stochastic software reliability model with imperfect-debugging and change-point. J Syst Softw 66(2):135–141
Singh J, Singh O, Kapur PK (2015) Multi up-gradation software reliability growth model with learning effect and severity of faults using SDE. Int J Syst Assur Eng Manage 6:18–25
Singh O, Kapur PK, Shrivastava AK, Mishra G (2017) A multi release cost model in distributed environment. Int J Reliab Qual Saf Eng 24(01):1750001
Singh O, Kapur PK, Shrivastava AK, Das L (2014) A unified approach for successive release of a software under two types of imperfect debugging. In: Proceedings of 3rd international conference on reliability, Infocom technologies and optimization. IEEE, pp 1–6
Singhal S, Kapur PK, Kumar V, Panwar S (2023) Stochastic debugging based reliability growth models for Open Source Software project. Ann Oper Res 1–39
Tamura Y, Yamada S (2006) A flexible stochastic differential equation model in distributed development environment. Eur J Oper Res 168(1):143–152

Tamura Y, Yamada S (2010) Performance evaluation of reliability assessment method based on stochastic differential equation model for a large-scale open source solution. Int J Syst Assur Eng Manage 1:324–329
Tamura Y, Yamada S (2014) Optimization analysis based on stochastic differential equation model for cloud computing. Int J Reliab Qual Saf Eng 21(04):1450020
Tamura Y, Yamada S (2017) Open source software cost analysis with fault severity levels based on stochastic differential equation models. Life Cycle Reliab Saf Eng 6:31–35
Tamura Y, Yamada S (2021) Performance assessment based on stochastic differential equation and effort data for edge computing. Softw Test Verif Reliab 31(6):e1766
Tamura Y, Yamada S, Kimura M (2006) Software reliability modelling in distributed development environment. J Qual Maint Eng 12(4):425–432
Tanaka H, Yamada S, Osaki S, Kawakami SI (1992) Software reliability growth model with continuous error domain—application of a linear stochastic differential equation. Electron Commun Japan (Part III: Fund Electron Sci) 75(3):37–46
Wang L, Hu Q, Liu J (2016) Software reliability growth modelling and analysis with dual fault detection and correction processes. IIE Trans 48(4):359–370
Xie M, Hu QP, Wu YP, Ng SH (2007) A study of the modelling and analysis of software fault-detection and fault-correction processes. Qual Reliab Eng Int 23:459–470
Xie R, Qiu H, Zhai Q, Peng R (2022) A model of software fault detection and correction processes considering heterogeneous faults. Qual Reliab Eng Int
Yaghoobi T (2020) Parameter optimization of software reliability models using improved differential evolution algorithm. Math Comput Simul 177:46–62
Yamada S, Nishigaki A, Kimura M (2003) A stochastic differential equation model for software reliability assessment and its goodness-of-fit. Int J Reliab Appl 4(1):1–11
Yang J, Liu Y, Xie M, Zhao M (2016) Modelling and analysis of reliability of multi-release open source software incorporating both fault detection and correction processes. J Syst Softw 115:102–110
Yang J, Zhao M, Chen J (2022) ELS algorithm for estimating open source software reliability with masked data considering both fault detection and correction processes. Commun Stat Theory Meth 51(19):6792–6817
Zhang X, Teng X, Pham H (2003) Considering fault removal efficiency in software reliability assessment. IEEE Trans Syst Man Cybern Part A: Syst Hum 33(1):114–120
Zhang X, Yang J, Wang W (2017) A stochastic differential equation-based software reliability growth model considering the effect of multiple fault removal modes. J Syst Softw 131:254–262
Zhu M, Pham H (2018) A two-phase software reliability modelling involving with software fault dependency and imperfect fault removal. Comput Lang Syst Struct 53:27–42

Chapter 10
Human Factors Engineering, Product Development and Sustainable Performance in Organizations: Issues and Challenges from an International Perspective

Pradip Kumar Ray

Introduction

The term 'Human Factors Engineering' (HFE) is considered a synonym of the term 'Ergonomics' in today's industrial and organizational context. While researchers and practitioners apply the principles of HFE to design workstations- or worksystems-related components and their interactions, the principles of ergonomics are equally emphasized and applied, as fundamentally these principles are to be used simultaneously while ensuring that the design of products, processes, or systems are made for human use.

The core issue in such a design endeavor is to consider both body and mind of a human working in a worksystem or at a workstation, with 'ergonomics' dealing with the 'body' part and 'human factors engineering' or 'human factors', in short, dealing with the 'mind' part of a human. It is assumed, with evident scientific reasoning, that both these parts are interrelated and hence, quality of interaction or interface may significantly affect the performance of humans in a worksystem. An acceptable and good quality interface design invariably ensures work methods and jobs or tasks to be undertaken by humans most comfortably and conveniently with adoption of safe practices and performance assurance. The term 'ergonomics', used in a Polis newspaper way back in 1857 and coined by Wojciech Jastrzebowski, is literally meant as 'law of work' ('ergo' means work and 'nomos' means law in Greek language) and primarily applicable to the body part of humans or physical activities being carried out related to jobs or tasks (Helander 2006; Bridger 2009). While we design jobs or

P. K. Ray (✉)
Department of Industrial and Systems Engineering, Vinod Gupta School of Management, Indian Institute of Technology Kharagpur, Kharagpur, West Bengal 721302, India
e-mail: pkr@vgsom.iitkgp.ac.in

P. V. Varde et al. (eds.), *Advances in Risk-Informed Technologies*, Risk, Reliability and Safety Engineering, https://doi.org/10.1007/978-981-99-9122-8_10

tasks in respect of these activities, we primarily focus on application of principles and knowledge in three specific areas or disciplines, viz work physiology, biomechanics, and workstation design (Helander 2006). These activities are referred to as 'below-the-neck' activities as carried out by humans. Application of ergonomics started in the UK industrial systems in 1930s.

While the jobs or tasks are being carried out, the 'mind' part also needs to be considered as activities are also controlled by human brain. Hence, the activities of the brain is referred to as 'above-the-neck' activities (Wogalter et al. 2001). The HFE is primarily focused on these activities. There has been substantial use of the HFE-related principles since mid-1950s in the USA-based industries. HFE is based as application of principles and knowledge in three areas or disciplines, viz. experimental psychology, system design, and human performance management. Today for all practical and design purpose, HFE and Ergonomics are used interchangeably, In fact, international societies of both have been merged into one.

Problems in Ergonomics and HFE at Workplaces

In today's industrial and organizational context, ergonomic/HFE-based approaches need to be used to address various kinds of problems focusing on to what extent the products, processes and systems are designed and implemented for humans at work (Hendrick 1996). Obviously, irrespective of technology being used, 'interaction or interface design' (human-product, or human-process, or human-system interface) has become a key issue in worksystem design. There are three basic components in a worksystem: human, machine, and environment; and against each one, there are three sub-components. Hence, there may be a number of interfaces to be considered for ergonomic design involving multiple sub-components. In majority of the cases and problems to be considered, designing of interfaces becomes a complex issue, and applicability of proposed design should be verified and validated with expectation of positive outcomes with health and fitness of humans at work remaining at an acceptable level and state. As appropriate interface design would always ensure a sustainable performance and job satisfaction of humans. In all likelihood, humans at work become 'intrinsically' motivated, a challenging and complex issue at the workplaces.

Basic Approaches in Worksystems Design: Ergonomics and HFE Perspectives

The traditional approach being followed for worksystem design in many manufacturing and production systems is based on 'FMJ' or 'Fit-Man-to-Job' philosophy. Since the beginning of operations and activities in factories and other systems, it has been assumed that, irrespective of the types and quality of jobs and worksystems, the humans are to be found and selected without bothering about the 'anthropometry' or characteristics of humans. In ergonomics or HFE-based approach, we need to follow a different approach, called FJM or 'Fit-Job-to-Man' for worksystem design. The basic principle to be followed is to design and allocate jobs to humans, at individual level, fitting with their anthropometry. If this principle is followed, there is guaranteed and reliable interface design, and hence, ergonomic/HFE-based performance of any worksystem, in all likelihood, is very high (Bridger 2009). While designing an interface, anthropometry of humans at work refers to specific human characteristics from job or task requirements perspective.

HFE/Ergonomics and Design of Worksystems

As has been highlighted, design of worksystems' interfaces is the primary goal of any study on ergonomic problems, their analyses and solutions. In this context, the specific objectives of such a study or research may refer to a number of relevant issues (Dray 1988). The important issues that need to be considered in almost all types of manufacturing and production systems, products and worksystems are as follows:

i. Interface design is to be considered at all phases of product development process, simultaneously or concurrently. The principles of ergonomics and HFE are most suitable and effective if concurrent or simultaneous engineering (C/SE) approach is followed.
ii. As far as practicable, a standard format is to be followed for ergonomic or HFE study for any worksystem involving data collection, data analysis, results and their interpretation, improvement plans and their implementation.
iii. As interface design may involve consideration of different types of sub-components and components in a worksystem, the interface components need to be classified. The classes that are considered critical problems are to be considered on a priority basis (Leaman 1980).
iv. In any interface design there may be anthropometric 'mismatches'. For improving the design of interfaces, the factors determining the level of mismatch are to be known and considered before acceptable condition of 'anthropometric match' is achieved in improved ergonomic design.
v. There are many kinds of standards and guidelines available in respect of different jobs and worksystem components and interactions. Against each of the jobs to be considered for ergonomic design analysis, these standards and guidelines are required to be followed, and new sets of guidelines and standards are to be developed for new kinds of workplaces as a long-term objective (Pheasant 1986; Sen 1984).

vi. As many of these factors. Particularly variations in anthropometry, may be related to basic racial features, social and living conditions of humans as well as the types of technology being used and adopted for jobs, measurement and evaluation of socio-technical implications has become a part of ergonomics/ HFE study, particularly in the context of macro-ergonomics (Sell 1980).

Ergonomic Problem Identification and Solutions

After shipping errors were reduced at a loading dock, the supervisor pointed with pride to his solution. A system of double and triple checks had been implemented for each load before the truck left the dock. When asked if the initial cause of the errors could have been the eight-digit product code used on the bill of loading, the stated reply was "There's no problem now," yet what went unsaid was "Don't bother me, I can solve my own problems." The value of the ergonomic solution was lost because this was perceived to be either a motivational problem or an unavoidable problem. By failing to recognize that it was a design-induced ergonomic problem, the appropriate solution was not implemented. As this example shows, a broad-based level of understanding and am ability to recognize problems is essential for each program to succeed.

The crucial hurdle to preventing and/or solving many industrial problems is simply their identification. Many serious problems are overlooked because they are not recognized as problems. Many examples like the one above abound in the real world.

The identification of problems is as much an art as it is a science. This chapter provides a wide range of ideas on the identification of problems. The identification of problems can be interactive with the client. In some cases, however, it may be preferable to identify potential problems on your own, prior to the involvement of the client. As you might expect, the techniques are dependent on situations.

Another dimension in the problem identification arena is the detection of potential problems rather than the identification of existing problems. This is an interesting situation since you may find yourself with the added burden of "selling the problem" in addition to developing and selling the solution to the problem.

The ultimate answer to the question of problem identification is simply to have people identify their own problems, without your intervention. By teaching others to identify problems (and eventually, how to correct those problems) you will have the best method of eliminating problems and of providing a safe and efficient operation.

The use of technological change can have both a positive and a negative impact on your work. It may allow correction of long-standing problems. At the same time, technology can cause problems that would not have appeared otherwise. The use of robots to solve lifting problems and the use of computer workstations that create glare and headaches are two examples that represent both sides of this coin (Dray 1988; Waters et al. 1993). The addition of technology in problem identification must not be overlooked.

Finally, the ergonomist must be able to evaluate the various problems and choose those with the largest positive benefit to work on. It is common to find one's self with more projects than time. There are some methods of evaluating the projects to determine the best approach to use.

Traditional Problem Identification

The traditional method of identifying problems is to look until you find something that is wrong. This hit-or-miss approach is simply not very effective. Unless one has had extensive experience, a checklist of some type can be valuable.

There a number of checklists available. Primary differences are that some of the checklists are more extensive than others and that some are designed for specific industries or special uses. Shorter checklists are valuable if one is making a cursory inspection or explaining ergonomics to others. It is possible to use one checklist as an example of the utility of designing a customized checklist to fit a particular need. Often, special checklists have been designed by trade associations or user groups. It is worthwhile to check with appropriate associations to see what they may have to offer in this area.

Checklists/Questions

There are many checklists designed to highlight ergonomics problems. A few examples are provided. The value of these checklists is to provide a complete list of potential problem areas so that no problems are overlooked. The checklist is a valuable tool for the less experienced ergonomist since it can direct the person to look for problems that have not yet been encountered. The checklist is also valuable for the experienced practitioner since it can serve as a reminder to prevent overlooking any areas.

The checklist is helpful in explaining to others the types of problems an ergonomist can help to solve. It is very useful during interactive introductory discussions as a way of explaining what will be expected during a walk-through inspection.

There are a number of different checklists, with a variety of uses. It is important to use this variety and flexibility to your advantage rather than letting it be a liability. Find the checklist that meets your specific needs rather than using the first one found. A checklist suited to evaluate the general workplace, task, and work environment is presented.

The checklist is designed specifically to solicit information from the personnel department or the line manager. It is very useful to use during interviews since it presents initial questions to determine problem areas and then uses detailed questions to diagnose the causes of these problems.

The checklist begins with symptoms (cues) that one may actually see in the work area and then leads to possible causes and solutions for that situation. It is effective when shared with the client in addition to being useful for the ergonomist.

Design and Development of Automated Systems for Ergonomic Evaluation of Human-Product Interfaces for Sustainable Performance Assurance

Background and Introduction

Healthcare industry is one of the most hazardous environments to work in. Employees in this industry are constantly exposed to a complex variety of health and safety hazards in the course of their work. Apart from physical hazards including exposure to radiation and light intensity, there are other ergonomic issues such as lifting and handling of patients including standing for long periods. Long working hours and shift work add to the stress of work. In a report published by the WHO, the disease burden caused by ergonomic issues among healthcare workers was found to be three million per year. The health care and social assistance sector is one of the largest service-providing sectors in the U.S. economy. It is projected to be the fastest growing sector through 2024, with healthcare occupations expected to add more jobs than any other occupational group. However, this sector also experiences over 550,000 nonfatal workplace injuries (4.1 per 100 incidence rate). Hence there is a need for an innovative system to take care of the on job difficulties and associated risk factors in the healthcare systems. The present proposal aims at an intensive approach of design, performance evaluation and validation in actual conditions.

Ergonomic Evaluation of Human-Product Interfaces

The main objectives of such evaluation are manifold: (i) To propose an assessment system for human product/process interface, (ii) To propose a methodology for developing an automated system for monitoring and control of the design interfaces, and (iii) To develop and apply such an automated system for a number of critical interfaces in the jobs involving healthcare and other worksystem.

The methodology consists of a number of interrelated steps, which are as follows:

Step-1: Model Ergonomic Risk

Some interventions call for developing and implementing communication technology models that promote meaningful communication and are successful for people with limited health literacy. These improvements in training healthcare providers in communication and patient education as well as the need to invest in a multilevel team approach to educate patients. However, the complex ideas that users of the healthcare system must understand, such as advance care planning, medication regimens and medical decision making, may not be optimally explained simply with verbal or written communication. Therefore these complex ideas can be facilitated by pictures, video, multimedia and other interventions aids beyond the written and spoken word.

1. Equipment and Data Collection

State-of-the-art wireless and wearable motion tracking and EMG devices will be used to objectively quantify postures throughout the procedure. The OpalTM system (APDM, Inc., Portland, OR USA) consists of 6 inertial measurement units (IMUs), each sensor containing accelerometer, gyroscope, and magnetometer.

These sensors have previously been validated and used for motion tracking in sports, spacesuits, and improving patient health. Prior to the workday, IMUs will be worn by the worker on his head, sternum, upper arms, and pelvis without interfering with performance. After donning the sensors and before starting each procedure, sensors will be calibrated using a static I-Pose as described by the manufacturer (NexGen Ergonomics, Montreal, Quebec). The units will log data at 64 Hz, onto an onboard memory card the same will be downloaded and processed after the full work day. Sensor data will be collected throughout the entire day, and task time will be defined as shift start to shift finish.

2. Data Analysis

Accelerometer, gyroscope, and magnetometer data from the IMU sensors may be processed into postural angles using scripts programmed in MATLAB® (R2015b, Mathworks Inc., Natick, MA USA). Specifically, these low-pass filtered (4th order Butterworth filter set at 32 Hz) data streams will be used to calculate neck flexion, torso flexion, and left/right shoulder elevation over time throughout the entire procedure with reference to the static I-pose. Neck flexion will be defined as the head motion relative to the torso in the sagittal plane, torso flexion will be defined relative to vertical, and shoulder elevation will be defined as the upper arm motion relative to vertical. These continuous computed postures will be summarized for each job into (1) mean posture angles, (2) range of motion, defined as the difference between 95th percentile and 5th percentile posture angles, (3) percent time in demanding postures, (4) percent time in static postures, and (5) number of posture changes per minute. Demanding postures will be based on previous commonly used definitions, >10° neck flexion, >20° torso flexion, and >45° shoulder elevation. Time in static posture will be defined based on previously published ergonomics literature as duration the flexion (neck and torso) or elevation (shoulders) angular velocities are <1 °/s normalized by procedure time. Posture changes will be defined as movements >10° with angular velocity faster than 1 °/s.

Step-2: Design of the Healthcare System with State-of-the Art Technology

1. Patient Monitoring

Some of the healthcare systems, the present proposal aims at developing are provides flexible and powerful patient surveillance through wearable devices at anytime and anywhere. A major challenge is to provide round-the-clock healthcare services to those patients who require it by wearable wireless medical devices. The application of wireless and mobile technologies has simulated a great advance in facilitating the development of electronic healthcare (e-healthcare). Because portable and wearable sensors can monitor a patient's health status in real time and automatically transmit the sensed data to patient healthcare management centres. Numerous portable devices are available that can detect certain medical conditions- pulse rate, blood pressure, breath alcohol level and so on- from a user's touch. Therefore all interventions of deployment of health information management through mobile devices introduce several challenges: data storage and management. It can be a mobile phone or a microcontroller platform capable of communicating with the internet. It also forwards information about the status of the sensors (e.g. proper operation, power source levels etc.).

2. Musculoskeletal Injuries (MSIs)

Despite numerous engineering and administrative controls that have been put in place, musculoskeletal injuries (MSIs) persist as the leading category of occupational injury in healthcare. Healthcare workers are reported to sustain MSIs at a rate exceeding that of workers in other industries. An evaluation of the risks, causes and activities associated with MSIs that includes all workers in healthcare is warranted as the interventions need to be specifically designed to target each group. Using advanced instruments is essential to characterize the burden of MSIs among workers in healthcare for the purpose of identifying appropriate prevention strategies and alleviating this significant burden.

Step-3: Interventions to Reduce Injuries

There are different types of injury prone tasks that are performed in the healthcare environment. Lower limb injuries are the most common injuries among sport practitioners and they are considered the main cause of sport practice disability. The first step when assessing the injury risk is the identification of the mechanism of injury, i.e. how the injury occurs. For that reason, in this work we will focus on studying the injuries produced in this way. Non-contact mechanisms refer to the injuries suffered when executing a specific movement that leads to a damage of the body, without involving any impact with an external element. A widely used approach is to reproduce the movements that provoke the injury in a clinical setting, without risk of harm, and measure selected features of the movement which are related to injury, known as injury risk factors. The analysis of these measurements can help to identify the high injury risk object. We shall however limit, due to paucity of time frame, our efforts to develop lower limb injury only.

Conclusions

In this talk, a number of human-factors engineering-based approaches with their application and the-state-of-the-art tools, techniques, and technologies to be used for ensuring sustainable performance for an organization are discussed. Both national and international scenarios related to issues and challenges being encountered by organizations with case examples are presented in the lecture.

References

Bridger RS (2009) Introduction to ergonomics, 3rd edn. CRC Press, Taylor and Francis

Dray SM (1988) From tier to pier: organizational adaptation to new computing architectures in designing a better world. In: Proceedings of 10th international congress of the IEA

Helander M (2006) A guide to human factors and ergonomics, 2nd edn. CRC Press, Taylor and Francis, Second Edition

Hendrick HW (1996) Road map to the future: revised strategic plan. HFES Bull 35(10):1–5

Leaman TB (1980) The organization of industrial ergonomics: a human-machine model. Appl Ergon 11:223–226

Pheasant ST (1986) Bodyspace. Taylor and Francis

Sell R-G (1980) Success and failure in implementing changes in job design. Ergonomics 13:809–816

Sen RN (1984) Application of ergonomics in industrially developing countries. Ergonomics 27:1021–1032

Waters TR, Putz-Anderson V, Garg A, Fine LJ (1993) Revised NIOSH equation for the design and evaluation of manual lifting tasks. Ergonomics 36:749–776

Wogalter MS, Dempsey PG, Hancock PA (2001) Defining ergonomics/human factors. In: Karwowski W (ed) International encyclopedia of ergonomics and human factors. Taylor and Francis

Chapter 11
Advancements in Safety Assessment Methods and Techniques for Analysis of Internal and External Hazards

Janaki Devi Kompella

Introduction

There are 438 nuclear power reactors operating in 32 countries with a combined capacity of about 390 GWe. With around 60 reactors currently under construction across 15 countries, the nuclear power programme is constantly expanding on a global scale (IAEA 2023). In parallel, plant modifications/upgrades and life extension programmes are being implemented, for instance in the USA and Europe to sustain the current generation capacity. About two-thirds of the reactors have been in operation for over 30 years, highlighting the importance of ageing management, Long Term Operation (LTO) and the need for new nuclear power capacity additions.

A considerable expansion is foreseen in the coming years in response to the climate change effects as the governments and companies worldwide are pledging to decarbonize rapidly and achieve net-zero emissions by 2050–2070. Moving away from fossil fuels and boosting the capacity of clean energy sources like nuclear power, hydrogen, and renewable energy in the energy mix are effective ways to speed up the decarbonization process. Nuclear energy will inevitably play a key part in this process. In a recent study, the International Energy Agency (IEA) projected that installed nuclear capacity would more than double to 890 GWe with all types of power reactors, including the Generation III+ large NPPs, Generation IV advanced reactors and small modular reactors[1] (IEA 2019). More than 80 Generation III+ and Generation IV designs are at various phases of development and deployment across 18 countries (Advances in Small Modular Reactor Technology Developments 2022).

[1] These reactors are intended to accelerate decarbonization also through their non-electric applications such as industrial heating, hydrogen production, desalination, district heating, and so on.

J. D. Kompella (✉)
RELSAFE PRA Consulting Private Limited, Mumbai, India
e-mail: dkompella@relsafe.co.in

P. V. Varde et al. (eds.), *Advances in Risk-Informed Technologies*, Risk, Reliability and Safety Engineering, https://doi.org/10.1007/978-981-99-9122-8_11

Given the current scenario and high ambitions for nuclear power in the future, there is an essential need for realistic and convincing safety assessments. Together with deterministic safety assessments, the nuclear sector is increasingly adopting probabilistic safety assessments and risk-informed decision making (RIDM) process more when making decisions about design, operation, maintenance. International organizations like the IAEA, USNRC, EPRI, ASME, OECD, NEI, et al. have brought out several methodological and data advancements in the recent years to enhance the quality of safety assessments.

2010–2022 has been the era of advancements in safety assessments methods and data. The Fukushima event reinforced the regulatory requirements on nuclear safety, necessitated design modifications and required the operating NPPs to perform safety re-assessments across several areas including seismic safety, long term station blackout, severe accident management, etc. With the increasing trend of climate change events such as hurricanes, heatwaves, droughts, external fires, etc., as well as impetus towards LTO, the safety regulations and requirements continue to evolve.

With the increasing interest in next generation reactors, the safety assessment process requires a fresh overhaul in light of several factors such as novelties in design (type of fuel, coolant, component), novelties in construction (multi-module setup vs integrated setup, buried nuclear island, etc.), limited operating experience, etc. which can influence the outcomes of safety assessment. These areas call for further improvements in methods, data and tools. International organizations are now developing new methodologies to address these special areas.

This paper discusses the reflections of application of latest methodologies and data in the areas of internal and external hazards analysis, including internal fire, internal flooding, seismic events, seismically induced fire and flood assessments, and the benefits brought out by these advancements. The paper also touches upon the special aspects associated with next generation reactors and endeavours taken by various international organizations to support the licensing and safety demonstration of these reactors.

Advancements in Internal Hazards Analysis

NPP regulators continue to place a greater emphasis on internal hazards, in particular internal fire and flooding events as they have the potential to induce multiple initiators and impact multiple safety systems thereby contributing significantly to the risk.

In the early years, fire and flood safety relied on engineering and design considerations such as the use of fire-and flood-resistant materials in construction and fire suppression/flood mitigation systems. In the 1980's–1990's, the nuclear industry recognized the need for more comprehensive safety assessment techniques and developed specific methodologies to evaluate risks associated with internal fires and floods. In parallel, the experience gained from notable events like the Browns Ferry fire incident and flooding events in Surry unit 2 and Palo Verde unit 1 introduced major changes to fire and flood protection requirements, safety culture and the scope of risk assessment.

For instance, the US NPPs developed dedicated PSAs for internal fires as part of Individual Plant Examination of External Events (IPEEE) program in the 1990's. These studies were based on EPRI fire induced vulnerability evaluation methods (Electric Power Research Institute (EPRI) 1992) and fire PRA implementation guide (EPRI 1995). Based on the insights and experience gained from pilot applications, the NRC and EPRI jointly developed a comprehensive methodology for internal fire probabilistic risk assessment (fire PRA), NUREG/CR-6850 in 2005 (United States Nuclear Regulatory Commission (USNRC) 2005) and supplement 1 (FAQs) in 2010 (United States Nuclear Regulatory Commission (USNRC) 2010). Likewise, EPRI developed a dedicated methodology for internal flood PRA, EPRI 1019194 in 2009 (Electric Power Research Institute (EPRI) 2009). These are still used widely as the state-of-the-art methodologies for performance of internal fire and flooding PRAs.

As the nuclear industry accumulated more and more operational experience (OPEX), utilities started to employ PRAs more when making decisions about design, operation, maintenance, and risk-informed applications. This warranted the need for more realistic data and methods for risk assessment (and in particular for hazards as they can mask the overall risk estimates by conservative assumptions). Over the years 2010–2020, significant data collection and analysis of OPEX allowed better estimation of generic fire ignition frequencies and pipe rupture frequencies for use in internal fire and flood PRAs (USNRC 2015; EPRI 2013; EPRI 2017). Also, understanding the fire/flood induced component failure modes allowed better characterization of the fire and flood impacts. Together with these, improvements in fire and flood propagation analysis methods, data, and advancements in computational capabilities enabled more realistic modelling of fire and flood scenarios. Modern fire and flood PSAs can thus take benefit of these developments to realistically estimate the risk from these hazards.

In recent internal fire and flood risk assessments performed for a European NPP, state-of-the-art methodologies, international best practice methods and data which evolved over the last 10 years, latest plant SSC databases, OPEX for fire and flood events, dedicated plant walkdowns, operator interviews for human reliability analyses, etc. were used. Further, common basis was established in deterministic and probabilistic assessments across different areas of the study viz: definition of fire and flood compartments, assessment of fire and flood induced impacts on equipment, fire and flood frequency estimation and detailed scenario analysis– as these areas drive the scope, approach, and analysis outcomes. The application of latest methodologies resulted in several benefits viz:

- Better characterization of full power and shutdown risks associated with fire/flood events.
- Precise identification of locations which are dominant from fire/flood risk point of view.
- More realistic fire/flood propagation calculations.

- Identification of important initiating events, and spurious actuations.
- Identification of important operator actions associated with event mitigation (and the need for additional scenario-specific operator actions).
- Re-verification of fire/flood compartment definitions (design basis).
- Design recommendations like installation of flood mitigation isolation valves, enhancement of cable fire protection, etc.

Following illustrations presents the salient highlights of the studies:

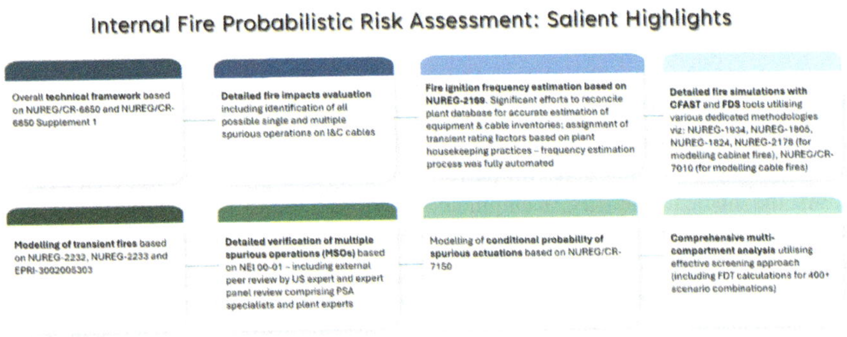

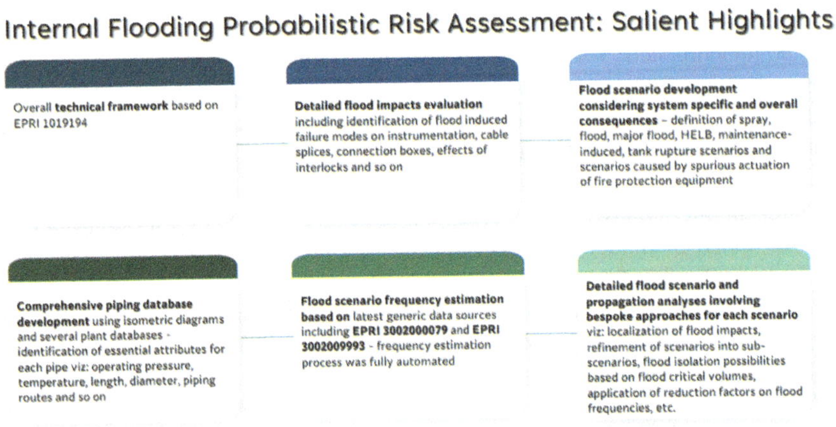

Deterministic Assessment of Internal Fire and Flooding Events: Salient Highlights

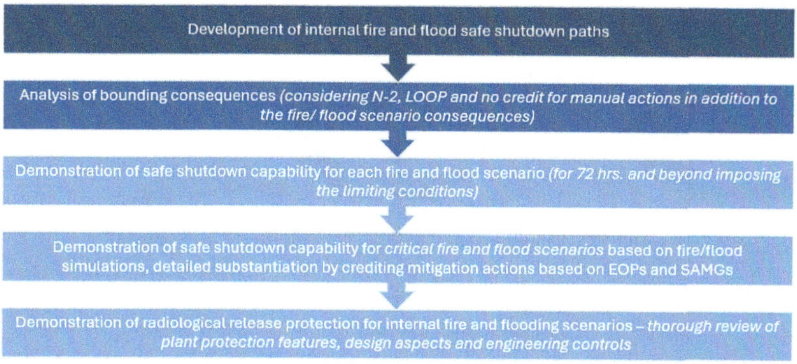

The data collection and analysis, understanding of fire/flood induced equipment failure modes, propagation mechanisms, structural failures, and human reliability analysis techniques continue to evolve in the industry.

Additional R&Dis ongoing to develop sophisticated computational models, simulation tools, and hydrodynamic models, which will allow for further accurate prediction and assessment of fire and flooding scenarios. Other areas of active research include concurrent and sequential risk modelling (i.e., risks stemming from concurrent or sequential fire and flooding scenarios), compounded impacts assessment (for instance determining the impacts of fire and flooding on structural integrity) and interdependency analysis (identifying, analysing, and modelling interdependencies between fire and flooding risks).

Advancements in External Hazards Analysis

External hazards analysis continues to gain importance due to their potential for widespread consequences. In the recent years, extreme weather events such as heatwaves, storms, hurricanes, external flooding, drought, heavy snowfall, etc. have become prominent due to climate change effects. Combination hazards such as seismically induced fire and flooding (SIFF) and other correlated hazards have also gained significant importance following Fukushima accident.

Considerable attention is paid to seismic events around the world as seismic risk still appears to be dominant in many studies. As a result of the reinforced regulatory requirements, new information on seismic hazards (for instance, in the US and Europe), experience from recent significant earthquake events (such as Tohoku, Mineral Virginia, Le Teil, etc.), improvements in methods and data, seismic risk assessment is aiming to achieve more realism and maturity. The following illustration shows how seismic risk assessment methods and perception have changed through time.

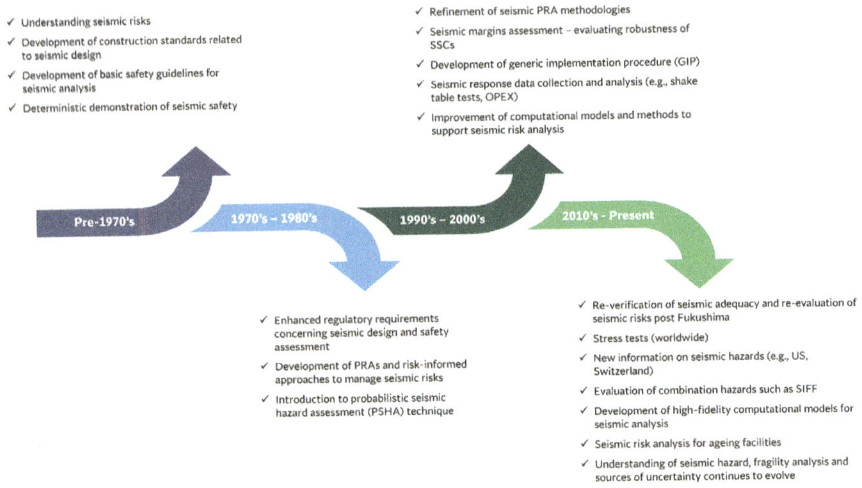

In a study conducted for one of the European NPPs, the scope of seismic safety assessment involved:

- Technical Accident Analyses to identify the SSCs important for success of safety functions from seismic safe shutdown paths (operational aspects and important insights from the full scope, multi-state PSA study were used for developing the seismic safe shutdown paths).
- Radiological Accident Analyses to identify the SSCs important from dominant radioactivity release scenarios.
- Development of comprehensive seismic equipment list (SEL), grouping of SSCs based on EPRI GIP classes (Seismic Qualification Utility Group (SQUG) 2001) (which was augmented with plant specificities) and assignment of initial correlation groups.
- Estimation of preliminary HCLPF and selection of representative (weaker) from each augmented GIP class for fragility analysis.
- Detailed verification of seismic adequacy and estimation of fragilities using Separation Of Variables (SOV) method.
- Deterministic demonstration of seismic adequacy for SSCs and safe shutdown capability.
- Probabilistic evaluation of seismic risk for a range of seismic hazard intervals (for all plant operating states).
- Re-verification of safety classification of SSCs based on functional approach (in line with IAEA SSG-30, IAEA, Vienna 2014).

Following are some of the salient benefits gained from the study:

- Effective identification of SSCs for a safety function (considering all support systems and interfaces relevant for the success of the function).
- Identification of dominant radiological release scenarios in the plant.

- Systematic identification of system and component boundaries for verification of seismic adequacy.
- Improvised granularity in equipment grouping to select candidate equipment for fragility analysis (while also covering diverse equipment classes).
- Selection of weaker SSCs that govern the robustness of seismic shutdown paths.

Seismically induced secondary hazards such as SIFF have gained greater attention in the recent times, as they proved to be of higher likelihood than originally assumed in the plant design. In a recent study performed for one of the European NPPs, a systematic and comprehensive assessment of SIFF events was performed based on the state-of-the-art EPRI methodology (EPRI 2018), covering both deterministic and probabilistic aspects in an integrated manner. This was a first of a kind application, as the probabilistic methodology was adapted to suit the deterministic framework. The data and analysis from the latest fire and flood deterministic and probabilistic studies were used as a starting point for SIFF analysis. Following are the salient highlights of the study:

- Application of a robust top-down screening process using qualitative and quantitative criteria (e.g., walkdown observations, seismic fragilities, risk measures) to identify the fire and flood sources of interest.
- Deterministic demonstration of plant safe shutdown capability for unscreened SIFF events.
- Quantitative evaluation of the additional risk impact due to SIFF using PSA model.
- Analysis of risks associated with explosive materials and seismic interactions.
- Independent international peer review of the study for technical validation on application of EPRI methodology (to reinforce the approach and methodological assumptions of this study, a review by TEPCO was also performed).

Challenges in Safety Assessment of Small Modular and Advanced Reactors

Nuclear reactor designs continue to evolve with variety of innovations, ranging from improvising the existing reactor designs to radical changes in overall design such as use of novel fuels, coolants, passive systems, and so on. While the safety aspects of the new reactor designs can be evaluated using available safety standards and good practices, new methods are required to address the specificities posed by advanced reactors—in terms of design uniqueness and modularisation aspects. In general, the challenges are related to:

- Lack of knowledge and OPEX of these reactors and potential safety impacts.
- High uncertainties due to novelties in design.
- Limited knowledge of initiating events, accident progression and physical phenomena.
- Defining acceptance criteria and safety margins for safety demonstration.

- Computer codes for thermal hydraulics and severe accident modelling.
- Choosing appropriate risk metrics for PSA (for e.g., Core Damage Frequency may not be the appropriate risk metric for molten salt reactor designs).
- Component reliability data (novel components, no OPEX, etc. New approaches are required for failure rate estimation).
- Passive system reliability analysis.
- Common cause failure analysis (for e.g., more shared systems in multi-module NPP design).
- Treatment of external events, software and digital system failures, human reliability analysis, and so on.

A tailored approach is thus required for each reactor design to address the novelties and specificities of the design. The nuclear industry has recognized these challenges, and international organizations like the IAEA, USNRC, EPRI and ASME have initiated several programs, published standards and guidelines to support the licensing and safety demonstration of next generation reactors. Some notable developments are:

- ASME PRA standard for non-LWR designs (ASME 2021)
- USNRC regulatory guide 1.247 endorsing the ASME non-LWR PRA standard (USNRC Regulatory Guide 2022)
- NEI peer review procedural guide (ANS Advanced Non-LWR PRA Standard 2020)
- EPRI advanced nuclear technology program (Advanced Nuclear Technology Program)
- Evaluation of risk analysis methods and tools for advanced reactors, by EPRI (EPRI 2023)
- IAEA's extensive review of the applicability of safety standards and guides for SMRs and advanced reactors (International Atomic Energy Agency 2022), to identify gaps and areas needing further development
- IAEA draft guide on safety demonstration of innovative technology in power reactor designs (IAEA DS 539).

Conclusion

This paper discussed the current status of nuclear power programs globally and the role of nuclear energy in net zero transitions, importance of safety assessments as utilities continue to request life extensions and the increased interest to deploy next generation reactors. The paper discussed the salient advancements in safety assessment domain—with focus on internal hazards such as internal fire and flooding, and external hazards such as seismic events. It also discussed how synergies can be established between deterministic and probabilistic safety assessments to support a balanced risk-informed decision-making. The paper also emphasised the growing importance of combinational hazards analysis such as the seismically induced fire and flood assessments (SIFF) and other external hazards such as droughts, external flooding, hurricanes, etc. induced by climate change effects. The paper also touched upon the challenges associated with safety demonstration of small modular and advanced reactors, and the endeavours taken by international organizations like the IAEA, USNRC, EPRI, ASME et al. in this regard.

References

American Society for Mechanical Engineers (ASME)/American Nuclear Society (ANS) RA-S-1.4-2021, Probabilistic Risk Assessment Standard for Advanced Non-Light Water Reactor Nuclear Plants

Electric Power Research Institute (EPRI) (1992) Fire-Induced Vulnerability Evaluation (FIVE), EPRI TR-100370

Electric Power Research Institute (EPRI) (1995) Fire PRA Implementation Guide, EPRI TR-105928

Electric Power Research Institute (EPRI) (2009) Guidelines for Performance of Internal Flooding Probabilistic Risk Assessment, EPRI-1019194

Electric Power Research Institute (EPRI) (2013) Pipe Rupture Frequencies for Internal Flooding Probabilistic Risk Assessments, EPRI-3002000079 (Rev. 3)

Electric Power Research Institute (2017) Pipe Rupture Frequencies for Internal Flooding Probabilistic Risk Assessments: Technical Update for High-Energy Line Piping, EPRI-3002009993, Revision 3, CA

Electric Power Research Institute (EPRI) (2018) Methodology for Seismically Induced Internal Fire and Flood Probabilistic Risk Assessment, EPRI-3002012980

Electric Power Research Institute (EPRI) (2023) Advanced Nuclear Technology: Evaluation of Risk Analysis Methods and Tools for Advanced Reactors, EPRI-3002026495

EPRI Advanced Nuclear Technology Program 41.08.01. https://www.epri.com/research/programs/065093

International Atomic Energy Agency (IAEA) (2014) Safety Classification of Structures, Systems and Components in Nuclear Power Plants, IAEA Specific Safety Guide No. SSG-30, IAEA, Vienna

International Atomic Energy Agency (IAEA) (2022), Advances in Small Modular Reactor Technology Developments - A Supplement to: IAEA Advanced Reactors Information System (ARIS), IAEA, Vienna

International Atomic Energy Agency (IAEA), Power Reactor Information System (2023). IAEA, Vienna. http://www.iaea.org/pris

International Atomic Energy Agency (IAEA), DS 539: Licensing Process of Nuclear Installations document preparation profile. https://www.iaea.org/sites/default/files/dpp539.pdf

International Energy Agency (2019) Nuclear Power in a Clean Energy System, IEA, Paris https://www.iea.org/reports/nuclear-power-in-a-clean-energy-system

International Atomic Energy Agency (2022) Title, Series Name Series Number [IAEA Preprint]. https://preprint.iaea.org/search.aspx?orig_q=reportnumber:IAEA-PC--8839

Nuclear Energy Institute (NEI) (2020) Performance of PRA Peer Reviews Using the ASME/ANS Advanced Non-LWR PRA Standard, NEI 20-09 (Rev. 0) draft

Seismic Qualification Utility Group (SQUG) and Electric Power Research Institute (EPRI) (2001) Generic Implementation Procedure for Seismic Verification of Nuclear Power Plant Equipment (Rev. 3A)

United States Nuclear Regulatory Commission (USNRC) and Electric Power Research Institute (EPRI) (2005) Fire PRA Methodology for Nuclear Power Facilities, Volume 2: Detailed Methodology, NUREG/CR-6850 / EPRI-1011989

United States Nuclear Regulatory Commission (USNRC) and Electric Power Research Institute (EPRI) (2010) Fire Probabilistic Risk Assessment Methods Enhancements, NUREG/CR-6850 Supplement 1 / EPRI-1019259

United States Nuclear Regulatory Commission (USNRC) (2015) Nuclear Power Plant Fire Ignition Frequency and Non-Suppression Probability Estimation Using the Updated Fire Events Database, NUREG-2169

United States Nuclear Regulatory Commission (USNRC) (2022) Acceptability of Probabilistic Risk Assessment Results for Non-Light-Water Reactor Risk-Informed Activities, draft guide for trial use, Regulatory Guide 1.247

Chapter 12
Integrated Approach to Nuclear Safety at NPCIL

Sameer Hajela

Introduction

The objective of Nuclear Power Plant Safety is to ensure and demonstrate that the risk from plant to public and plant personnel is acceptably low. The safety philosophy is ensured with both deterministic and probabilistic approach of safety assessment. The integrated decision-making process (referred to as a Risk Informed Decision Making (RIDM)) is a structured process in which all the insights and requirements relating to a safety or regulatory issue that needs to be dealt with, are considered in reaching a decision (IAEA 2011). A Risk Informed Decision-Making process, along with relevant regulatory and legal requirements, ensures that the defence in depth and adequate safety margins are maintained. Safety philosophy based on deterministic approach includes various layers, in line with defence-in-depth principle, large safety margins, prevention of common cause failure. A comprehensive Probabilistic Safety Assessment (PSA) (NAPS 2014) complement the traditional deterministic approach to facilitate more informed and balanced decision-making process.

At NPCIL, areas where PSA supports deterministic approach, include, events categorization using Defence in Depth, development of the Symptom Based Event Handling through a Computer Based System- Symptom Based Intervention Guidelines Management System (SIGMAS) and Accident Management Guidelines (AMG). Inline with these applications of PSA, it also gives Point-in-time risk by the Risk Monitor application which is an operator aid in decision making to plan the maintenance and the configuration control.

S. Hajela (✉)
Nuclear Power Corporations of India Limited Headquarter, Mumbai, India
e-mail: edrsa@npcil.co.in

P. V. Varde et al. (eds.), *Advances in Risk-Informed Technologies*, Risk, Reliability and Safety Engineering, https://doi.org/10.1007/978-981-99-9122-8_12

Risk-Informed Decision-Making

Risk is a combination of probability and severity of undesirable event while safety is freedom from unacceptable risk. Risk Informed Decision Making (RIDM) is a complementary utilisation of deterministic and probabilistic approach to satisfy safety goals. The quantitative and qualitative risk information provided by the Risk Monitor is used as one of the inputs into an integrated (risk-informed) decision making process at the Plant. Deterministic approach as the basis for making decisions on safety issues provides high level criteria to ensure defence in depth and adequate safety margins. PSAs have been developed and is increasingly being used to complement the deterministic approach. The risk informed approach aims to integrate in a systematic manner quantitative and qualitative, deterministic and probabilistic safety considerations to obtain a balanced decision. Figure 12.1 represents the various elements of RIDM process.

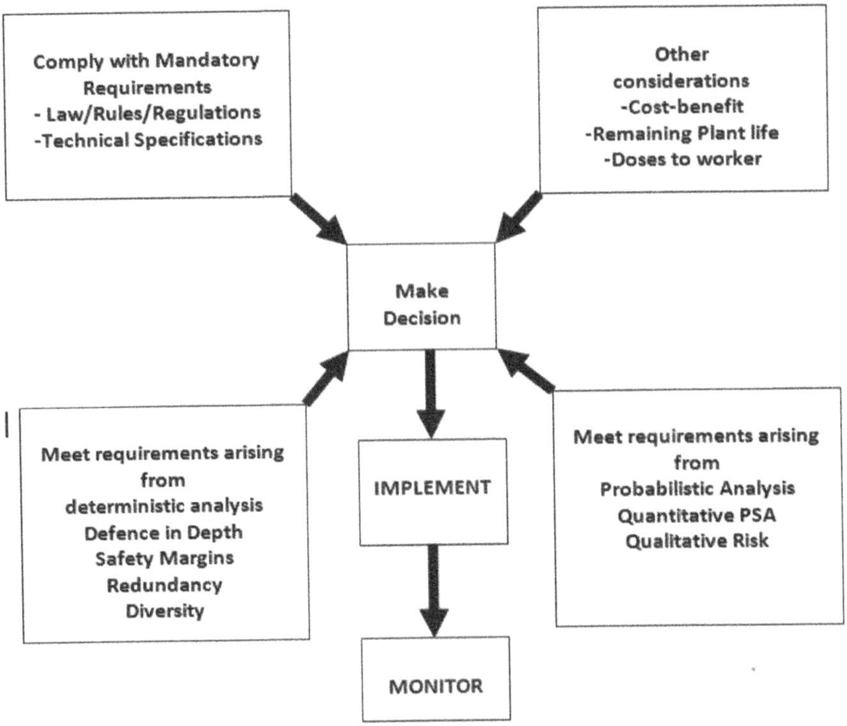

Fig. 12.1 Risk informed decision-making (IAEA-TECDOC-1436) (IAEA 2005)

Regulatory Requirements and Ridm

In India, Atomic Energy Regulatory Board issues Regulatory Code and Guide for licensing and regulation of Nuclear Power Plants. According to this Code and Guide, the licensee has to use PSA in support of licensing of new NPPs & Plant Modifications and back fitting of existing plants and for resolving safety issues at operating NPPs. According to regulatory consenting process, full scope Level-1 PSA needs to be submitted before first approach to criticality. During Periodic Safety Review to complement the Deterministic Assessment, consideration is given to Probabilistic Safety Assessment as an input to provide insight to relative contribution to safety of various aspect of plant. RIDM is being used for point in time risk assessment for maintenance planning and for technical specification modifications by control room operator through Risk Monitor. Target core damage frequency, large early frequency and system unavailability value is also needs to be satisfied (IAEA 2010; IAEA-TECDOC 2005).

PSA Quality Attributes for Risk-Informed Applications

A high quality plant specific PSA is being used in IPHWRs for PSA applications. These PSA are based on in house deterministic analysis of Anticipated Operational Occurrences (AOO), Design Basis Event (DBE) & Beyond Design Basis Events (BDBE). A full scope Level-1 and limited Level-2 PSA, addressing all internal hazards and external hazards like floods due to Tsunami, Storm Surge, Precipitation, Dam Break etc. as applicable, with plant specific reliability parameters are being used for PSA applications in IPHWRs (Anita et al. 2010; Kamyab et al. 2019).

Risk Monitor and Probabilistic Safety Assessment

Risk monitors are based on the Probabilistic Safety Assessment model, the essential difference is that the Risk Monitor is designed to be used by nuclear power plant operator, rather than a PSA specialists, the user not require to have specialist knowledge of PSA techniques. The PSA model needs to be amended to remove some of the simplifications for risk monitor model. In particular, initiating events such as LOCA that are modelled in the Living PSA as a lumped event in one of the coolant loops may need to be replaced by individual initiating events in each of the coolant loops; system alignments and the choice of running and standby trains of normally operating systems may need to be modelled explicitly; initiating events that have been screened out of the PSA may need to be reinstated if they could be significant in some plant (CSNI 2004).

Risk Monitor Implementation at IPHWRS

The Risk Monitor concept is implemented through Risk Spectrum@ Risk Watcher Software. The main objective of implementation of Risk Monitor is to produce a PSA tool to generate risk information for use in the day-to-day management of operational safety. A comprehensive full power risk monitor models are developed for IPHWRs and being used for operational risk management by control room operators. A limited scope risk monitor model is also developed for 700MWe IPHWRs shutdown mode of operations. To address the challenges in developing Risk Watcher Model due to dynamic plant configurations in shutdown mode of operations, an Initiating Event Model consisting of Fault Tree and Event Tree was developed. This Initiating Event Model gets configured implicitly as per plant operating state implemented in Risk Monitor. Decision making using a Risk Monitor usually requires the definition of three types of quantitative criteria—Risk Bands, Operational Safety Criteria and Allowed Configuration Time.

The colour coded risk bands are associated with the actions that need to be carried out by the plant operators (Fig. 12.2).

Operational Safety Criteria (OSCs) are specified to define the boundaries between these risk bands. OSCs are defined in terms of absolute risk levels of the baseline risk. For full power operation, the boundary between the regions of low and moderate risk has typically been set at about the level of the average risk (CDF or LERF) calculated in the PSA when maintenance outages have been taken into account. This is consistent with the aim of the plant operators who are trying to keep the plant risk down to a low level that is below the average risk and hence they would want to know when the risk is above this level. There is a broad international consensus that the boundary between the regions of high and unacceptable risk should be set at 10^{-4} per year for CDF and 10^{-5} per year for LERF. These numerical values relate to a full scope PSA as defined above and they would need to be adjusted if the scope of the PSA is less

Band Colour	Risk
Unacceptable	Unacceptable risk band which is not entered voluntarily and immediate action needs to be taken to reduce the risk.
High	High Risk Band where severe time restrictions need to be imposed and compensatory measures may be required
Moderate	Moderate risk band where maintenance needs to be completed quickly
Low	Low Risk Band where maintenance can be carried out with no restrictions

Fig. 12.2 Bands for operational safety criteria (NAPS 2018)

Risk Level	Degradation in D-i-D	Scenario
Unacceptable	Complete	Success criteria not met
High	High	Success criteria just met
Moderate	Minimal	At least one redundant system/train/sub-train/component available
Low	Nil	System/train/sub-train are completely available

Fig. 12.3 Qualitative Success Criteria and Bands (NAPS 12)

than this. The boundary between the regions of moderate and high risk is set at an intermediate value.

Allowed Outage Time (AOT) and Allowed Configuration Time (ACT) related to the maximum time for which a component/train unavailability or a plant configuration is allowed to persist before some action has to be taken to move the plant to a safer state.

The qualitative risk measures include coloured coded displays that indicate the status of safety functions/safety systems and their ability to respond to plant transients (Fig. 12.3). The level of defence-in-depth available for each of these safety functions is used as the basis for the qualitative risk measures displayed by the Risk Monitors. Defence in depth are defined for both operational and safety requirement by specifying success criteria (Varde and Pecht 2018).

Case Study: Risk Monitor at Naps

For operational requirement, success criteria are defined as per NAPS Technical Specification and for safety requirement success criteria are based on deterministic safety analysis of NAPS. Rules of qualitative risk measures as shown above are followed to display in colour codes, the degradation of a system due to unavailability of some of its components. For consistency, same rules are followed for both safety and operational requirements.

For NAPS, four risk bands based upon absolute risk level are defined. The boundary values were adjusted to include the anticipated risk other than internal initiating event as this Risk Watcher Model is not representative of Full Scope PSA for this Risk Monitor implementation (Fig. 12.4).

The Risk Watcher display with two DG sets under maintenance is considered to demonstrate the above criteria. It can be seen that in operational requirement of Class-III power supply is shown in 'RED" as technical specification allows outage of only one DG set. However, safety requirements of Class-III power supply is shown in "ORANGE" since one DG set just meets the success criteria for safety requirements. The operational requirement is "RED" while safety is in yellow since defence in depth

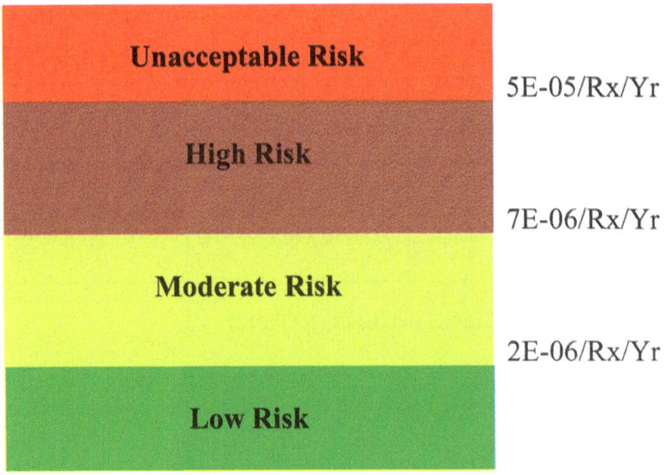

Fig. 12.4 Operational safety criteria for NAPS (2018)

based on safety requirements are still met. Figure 12.5 shows the impact on D-i-D for the above outage. The CDF (2.34E-05/yr.) for above outage is seen in "ORANGE" region as per the operational safety criteria for NAPS. The Fig. 12.6 displays the CDF history corresponding to the outages listed in Table 12.1.

PSA Support to Deterministic Analysis

With the help of PSA inputs, the contributors of the core damage events are identified and their accident progressions are studied. Among the studied events, the most governing scenario are taken as basis for developing accident management guidelines. Based on Level-1 PSA, different safety studies are performed to modify the design/change the logics for further enhancement of safety and improvement in Core Damage Frequency. Inputs from PSA studies are taken to arrive at various plant parameters that are used in Symptom Based Event Handling Scheme for PHWRs.

Conclusions

The considerable number of successful applications of PSA indicates the strength of probabilistic approach in decision making process. Implementation of Risk Monitor in NPCIL is arguably the most influential development in area of PSA applications. Risk Monitor is being used for planning maintenance activities to ensure that high peaks in the risk are avoided wherever possible and average risk is maintained as reasonably low. PSA results are also being used in NPCIL to develop SIGMAS and

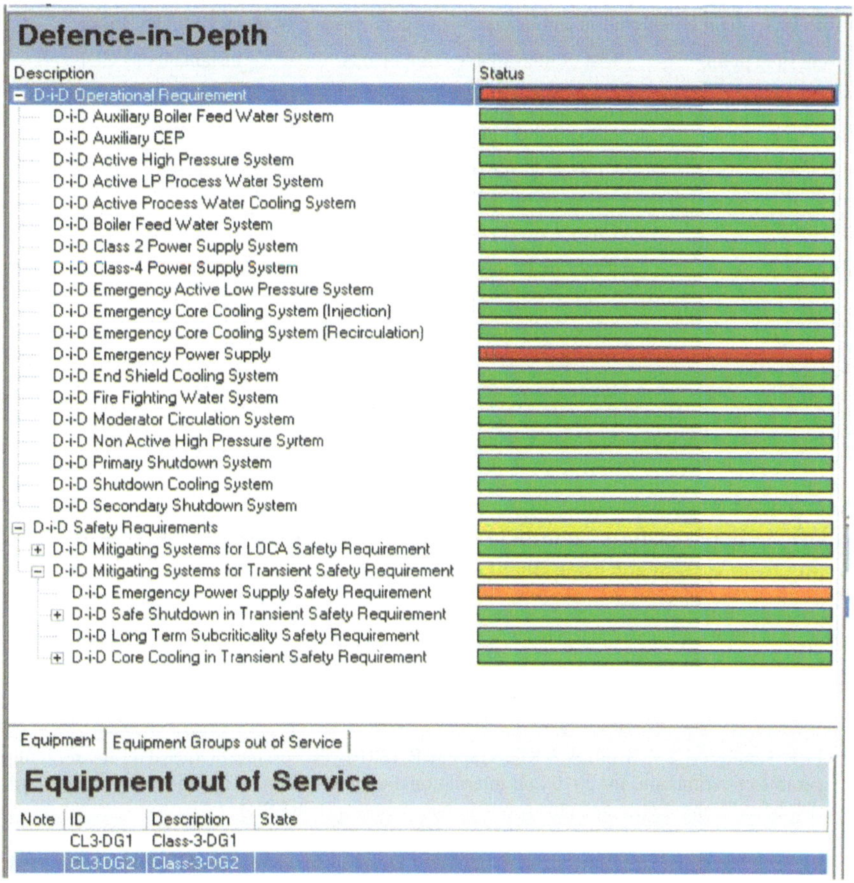

Fig. 12.5 Impact on D-i-D due to outage of diesel generator-1 and 2 (NAPS 2018)

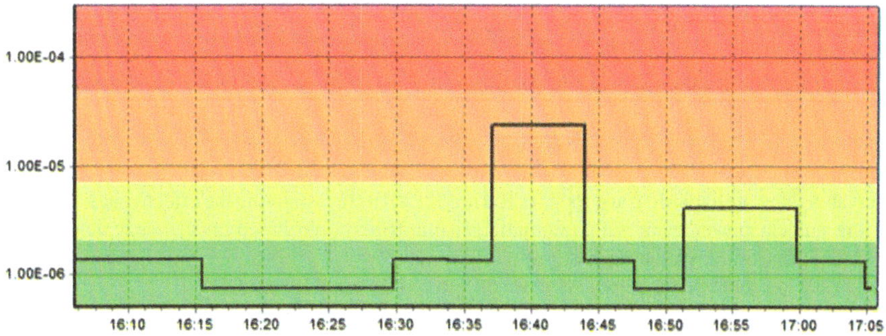

Fig. 12.6 Risk graph (CDF) (NAPS 2018)

Table 12.1 Plant configuration and instantaneous core damage frequency (NAPS 2018)

Event time point	List of unavailable components	Instantaneous CDF
16:29:49	Class-3-diesl generator-1	1.32E-06
16:33:51	Class-3-diesl generator-1	1.31E-06
16:37:08	Class-3-diesl generator-1 Class-3-diesl generator-3	2.34E-05
16:43:57	Class-3-diesl generator-1	1.32E-06
16:51:21	Fire wire pump-3 Class-3-diesl generator-1	4.10E-06
16:59:33	Class-3-diesl generator-1 Emergency core cooling pump-1	1.32E-06
17:04:00	Emergency core cooling pump-1 Emergency core cooling pump-2	7.39E-07

Accident Management Guidelines. A systematic integrated use of PSA and Deterministic Safety Analysis contributed to reduction in risk of operation of IPHWRs arising from system and component and human failure events and their interactions.

References

Anita P, Reddy ALVV, Srinivas VKG, Guptan R (2010) Processing of raw data-for reliability parameter estimation. In: 2010 2nd international conference on reliability safety and hazard - risk-based technologies and physics-of-failure methods (ICRESH)

Construction Innovation: Information, Process, Management, vol 14, issue 2 (2014)

CSNI. Risk Monitors-The State of the Art in Their Development and Use at Nuclear Power Plants [R]. France: OECD NEA/CSNI/R (2004)20

Development of a model for Risk informed decision making for equipment Outage management on "risk watcher "Software: implementation for NAPS-1&2, NAPS-1&2/DN/01560/00001/R0 (2018)

IAEA-TECDOC-1436-Risk informed regulation of nuclear facilities: Overview of the current status (2005)

IAEA—SSG-3- Development and Application of Level 1 Probabilistic Safety Assessment for Nuclear Power Plants (2010)

IAEA—INSAG-25—A Framework for an Integrated Risk Informed Decision Making Process (2011)

Kamyab S, Nematollahi M, Yousefpour F, Karimi K (2019) Investigating the effectiveness of outfitting nuclear power plant with automatic seismic trip system based on the early detection of earthquake. Soil Dyn Earthq Eng

NAPS (2014) Probabilistic Safety Assessment Level-1(Internal Events) Rev-1

Varde PV, Pecht MG (2018) Risk-based engineering. Springer Science and Business Media LLC

SPRINGER NATURE

GPSR Compliance

The European Union's (EU) General Product Safety Regulation (GPSR) is a set of rules that requires consumer products to be safe and our obligations to ensure this.

If you have any concerns about our products, you can contact us on ProductSafety@springernature.com

In case Publisher is established outside the EU, the EU authorized representative is:

Springer Nature Customer Service Center GmbH
Europaplatz 3
69115 Heidelberg, Germany